PALO DURO VISTAS

PALO DURO VISTAS

A Journey Through Palo Duro Canyon

By

William MacLeod

TEXAS GEOLOGICAL PRESS

ALPINE, TEXAS

Copyright © 2007 by William MacLeod

All rights reserved

First Printing, July, 2007

Printed and Bound in the United States

Cover Photo: "Fortress Cliff"
by William MacLeod

Publishers Cataloging in Publication Data

MacLeod, William

    Palo Duro Vistas: a journey through

    Palo Duro Canyon/ by William MacLeod

    p. cm.

    ISBN 978-0-9727785-2-7

    1. Geology – Palo Duro Region (Tex.). I. Title.

QE168.P28 M24 2007

557.64

TEXAS GEOLOGICAL PRESS
P.O. Box 967
Alpine, Texas 79831
www.texasgeologicalpress.com

For Martha

This view, taken from the Sorenson Guest Cabin next to the Interpretive Center, looks across Timbercreek Canyon to the Spanish Skirts in the center of the photograph.

PALO DURO VISTAS

Introduction

Palo Duro Canyon is one of the most photogenic places in Texas. The wonderful array of subtle hues from the rock layers contrasted against the lush vegetation and bright red shale along the canyon floor creates a setting that is unmatched in the southwest.

The narrow gorge, protected from the prevailing west winds and the bitter winter northers, has been a place of shelter for humans from their first arrival in the Panhandle, 12,000 years ago. Water and grass supported abundant wild life for hunter-gatherers who continued to use the canyon until the Federal government drove them into reservations in the late nineteenth century.

Today, Palo Duro Canyon is one of the best known landmarks in Texas, a tribute to the industry and vision of local people, who have publicized the canyon over the years, especially through the annual summer pageant "Texas". The play recounts the history of the region in dramatic reenactments. Visitors can hear the ghosts of Texas history in the canyon, the Paleo-Indians who arrived near the end of the Ice Ages, the Comanches who took control over the area in the eighteenth century, and Charles Goodnight and the early cattlemen who settled in the area in the 1870s.

Palo Duro Vistas takes you on a journey from the city of Canyon down into Palo Duro Canyon, describing and illustrating the terrain and the geology along the way. Although the canyon area is only a small section of Texas, the geology along the route encompasses several interesting facets: the final retreat of the ocean with the near-extinction of life at the end of the Permian period, the rebirth of life in the upper Triassic, the development of the extraordinary Ogallala Formation, and the wind-blown sands of the most recent two million years with the amazing fossils found in them.

The final section of the book briefly recounts the Coronado expedition of the sixteenth century, exploration of the Llano Estacado in the nineteenth century, and the story of Charles Goodnight from his early days to his settling in the canyon in 1876.

For those who want to delve more deeply into the history and geology of the Park, I have included notes at the end of the book where my principal sources are identified.

The Lighthouse

Photograph © Texas Parks and Wildlife Department.

PALO DURO VISTAS

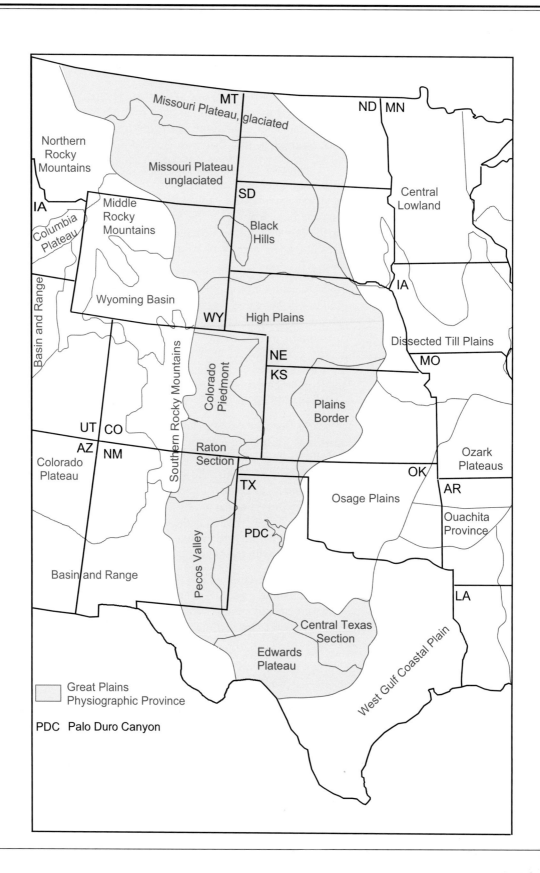

The Canyon Setting

Palo Duro Canyon is in the Great Plains physiographic province of the United States. Physiographic provinces are based on type of terrain, type of rock, and geological structure and history. The current classification is shown on the opposite page, with the Great Plains outlined in yellow. The Great Plains is further subdivided into subprovinces, such as the High Plains, the large mesa or tableland of grass prairies stretching from South Dakota to west Texas and rising from several hundred to more than a thousand feet above the country around it.

Palo Duro Canyon is cut into the Southern High Plains or Llano Estacado, as the section of the subprovince south of the Canadian River is called. The Southern High Plains cover an area of about 30,000 square miles, sloping to the southeast from an altitude of 5,000 feet above sea level at its northwest corner in New Mexico to less than 3,000 feet in the southeast (see contour map on page 13). At the entrance to Palo Duro State Park, the surface elevation is 3,439 feet.

The Llano Estacado shows up conspicuously on the shaded relief map (page 12), bounded by the Canadian River valley to the north and escarpments to the east and west. To the south, it blends into the Edwards Plateau at Johnson Draw just south of Midland.

The eastern escarpment is sharp, with Palo Duro Canyon the longest and deepest canyon to penetrate it. The tributary of the Red River that flows down Palo Duro Canyon is about 600 feet below the canyon rim at the State Park but descends more quickly than the slope of the upper surface as it goes downstream, so that by the time it reaches the confluence with Tule Canyon, the canyon is about 1,000 feet deep. The western escarpment is less impressive. At its highest, at the Mescalero Escarpment, it is only about 400 feet high, although about 900 feet above the Pecos River to its west. By the time the escarpment crosses Interstate 20, about 50 miles west of Midland, near its southern border with the Edwards Plateau, it is only about 80 feet high.

A great deal of research has been carried out into the underlying structures of the northern Llano Estacado because it was at one time considered as a possible site for burying nuclear waste. One of the research findings was that evaporites, minerals that precipitate from salt water bodies and are present in great thicknesses under the Llano Estacado, have been partly dissolved and carried away in ground-water east of the Caprock escarpment and under the Canadian and Pecos river valleys. For example, evaporites are 1,100 feet thick under the town of Canyon but only 700 feet thick under the Caprock escarpment and 400 feet thick in the Canadian River valley.

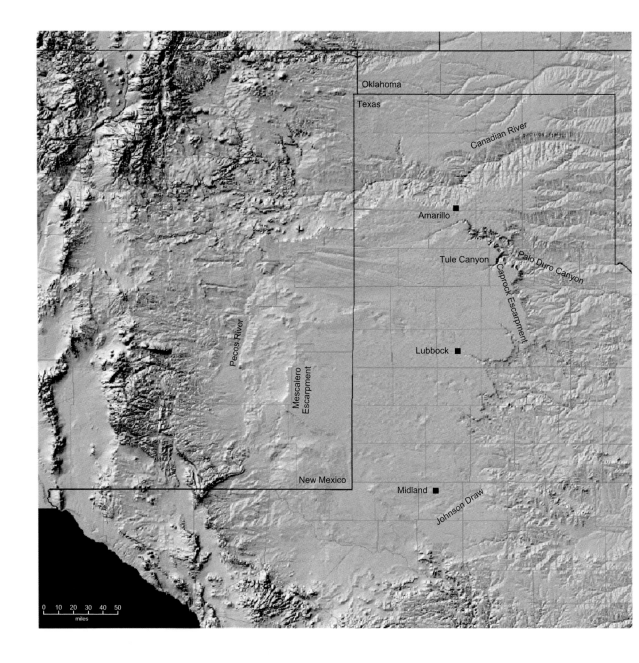

Shaded Relief Map of the Llano Estacado

To produce this "shaded relief" map, the U.S. Geological Survey compiled a data set, their National Elevation Dataset, of elevations every 10 meters (about 33 feet) in the United States and plotted it so that, by computer wizardry, it has been made to appear as a three dimensional representation with the sun shining across the terrain from the west. The advantage of this method is that topographic features show up with great clarity, even better than in satellite photographs.

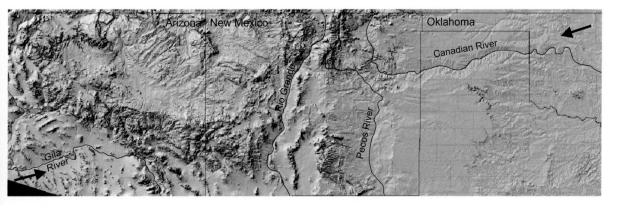

Above *The Socorro Fracture Zone, marked by the black arrows, can be followed from the Canadian River to the Gila River in Arizona.*

As the salts are carried away, the surface above subsides and slumps, creating broken ground. The shaded relief map opposite shows such areas of broken ground to the east of the Caprock escarpment and in the Canadian and Pecos River valleys. The "breaks", as such areas are called in north Texas along the Canadian River, are 25 miles wide and nearly 50 miles wide along the Pecos.

Ever since the onset of satellite mapping, observers have noted a prominent set of linear features such as rivers, depressions and faults from the Gila River in southwest Arizona to the Canadian. This is the 50 mile wide Socorro Fracture Zone, marked by the black arrows in the shaded relief map above.

The zone intersects the valley of the Rio Grande River about midway across New Mexico at the point where it widens dramatically (see map on page 40) and where a sizeable body of igneous rock lies 12 miles below the surface. This body arose along channels at the intersection of the fracture zone with the Rio Grande valley, which itself runs along a major rift in the Earth's crust.

Many recent major earthquakes, including several in the Texas Panhandle, have occurred along the fracture zone. Earthquakes occur when blocks of rock move against each other. In the Panhandle, the surface north of the Canadian River is some 250 feet lower than the one to its south (see map on next page), indicating that there has been substantial faulting along the Canadian River valley. Finally, the Canadian River follows the fracture zone across the Panhandle, turning sharply left to do so and the Pecos River also makes a detour east along the zone.

The Socorro Fracture Zone is probably a weak zone in the Earth's crust, one that has been there since the Precambrian era, more than 500 million years ago, and the fracturing of rocks along the zone has enhanced groundwater circulation along it, leading to slumping, along which the rivers follow.

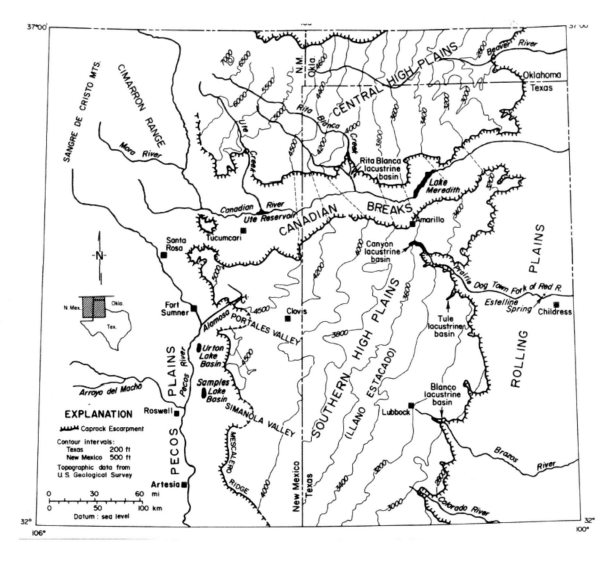

The Geology of the Palo Duro Area

The table on the following page gives the general geological time scale, with periods represented by formations in the Palo Duro area shown in red. Almost all the rocks around the canyon are classified as sedimentary, made from:

> rock fragments that are created by natural processes such as weathering, and that are transported by air, water or ice

> or from materials that accumulate by other natural processes such as precipitation from solution or secretion by organisms.

Sediments form loose layers on the Earth's surface such as silt, sand and gravel and over time become *indurated* or cemented together to form sedimentary rocks.

Above *Contour map showing the Southern High Plains sloping downhill from about 5,000 feet above sea level near Santa Rosa in New Mexico to 3,000 feet at the head waters of the Colorado River.*

Note: the surface north of the Canadian River is some 250 feet lower than the one to its south.

Right *Geological time scale; strata present in the Palo Duro area are shown in red.*

Ma = million years ago

Eon	Era	Period	Epoch	From Ma	To Ma
Phanerozoic	Cenozoic	Quaternary	Holocene	0.01	0.0
			Pleistocene	1.8	0.01
		Tertiary	Pliocene	5.3	1.8
			Miocene	23.8	5.3
			Oligocene	33.7	23.8
			Eocene	54.8	33.7
			Paleocene	65.0	54.8
	Mesozoic	Cretaceous	Upper Cretaceous	98	65
			Lower Cretaceous	144	98
		Jurassic		206	144
		Triassic		251	206
	Paleozoic	Permian		290	251
		Carboniferous	Pennsylvanian	323	290
			Mississippian	354	323
		Devonian		417	354
		Silurian		443	417
		Ordovician		490	443
		Cambrian		543	490
Proterozoic				2,500	543
Archean				4,600?	2,500

The ages of sedimentary rocks are determined by identifying the fossils they contain. In addition, some Palo Duro rocks contain volcanic ash beds made up of glass droplets derived from molten rock or *magma*. Ash is created by volcanic explosive eruptions and can be carried long distances by wind. These droplets are quite useful because their age can be calculated by measuring the ratios of certain isotopes of their elements.

Sedimentary rocks are classified by the size of their grains or *clasts*:

clay clasts are less than .00016 inch in diameter; if indurated, the rock is called mudstone; if it has laminations or partings it is called shale

silt clasts are between .00016 and .0025 inch in diameter; indurated, the rock is called siltstone

sand clasts are between .0025 and .08 inch in diameter; indurated, it is called sandstone

gravel clasts are greater than .08 inch in diameter; indurated, it is called conglomerate.

Other terms used in the descriptions in the following pages include:

alluvium, a general term for sand, silt, gravel etc. deposited by streams

loess, wind-blown dust

caliche, also called calcrete, a hard mass of calcium carbonate precipitated from water rising to the surface in dry periods of the year

calcereous, containing calcium carbonate

evaporite, a non-clastic rock formed by the precipitation of salts from water through evaporation, including gypsum $Ca(H_2O)_2(SO_4)$, anhydrite $(CaSO_4)$ and halite or common salt $(NaCl)$

playa, a closed basin underlain by stratified clay, silt or sand, occupied by an intermittent lake in the wet season.

Pages 18-19 show an area map and explanations of the colors used in the map, all taken from publications of the Bureau of Economic Geology in Austin. Pages 20-21 show a map of the section of the State Park you can see from the park road and explanations of the colors used, also based on a publication of the Bureau of Economic Geology. The State Park map is more detailed but the maps are otherwise consistent.

To summarize the geology of the area, there are two sets of strata in the area, one about 250 million years old and the other 11 million years old or less. The younger strata consists of sands, silts, clays and gravels of the Tertiary and Quaternary periods, identified on the map explanations by names such as Alluvium, Playa Deposits, High Alluvium, the Blackwater Draw Formation, the Ogallala Formation and the Cita Canyon Lake Beds.

The younger strata overlie shales, mudstones and sandstones of the Triassic Trujillo and Tecovas Formations (together called the Dockum Group on the area map) and shales, mudstones and gypsum beds of the Permian Quartermaster and Cloud Chief Formations.

The next section of the book describes the scenery and rocks on view along Highway 217 from the town of Canyon to Palo Duro State Park and along Park Road 5 down into the canyon to a turnaround and return along the Alternate Park Road 5.

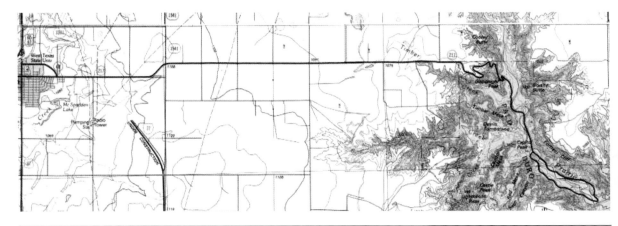

Taking a Trip through the Canyon

The road trip begins in the town of Canyon opposite the Plains Panhandle Historical Museum. The round trip to the end of the road in the canyon and returning to the museum is slightly over 40 miles. Be sure to reserve time to visit the museum, which is one of the best I have ever seen in Texas, and includes a splendid collection of Panhandle fossils.

The town of Canyon is situated on a finger of high ground between two creeks, to the north Palo Duro Creek, which downstream becomes the Prairie Dog Town Fork of the Red River, and one of its tributaries, Tierra Blanca Creek to the west and south. The high ground is underlain by deposits of silt, limestone and loess of the Blackwater Draw Formation.

South and east of Canyon lies a basin some 9 miles in diameter where the Blackwater Draw sediments have been eroded away and the underlying Ogallala Formation, shown on the regional map as a brown circular area, is at the surface.

Forty or more similar large basins are scattered over the Southern High Plains, most of them filled with salt water. They most likely developed as Permian evaporite beds dissolved in ground-water causing the surface to subside. The nearest Permian evaporite layer is 860 feet below the surface at Canyon.

Later, as the valley of the Red River Prairie Dog Town Fork eroded back to the north and the Tierra Blanca Creek valley developed, the lake was drained and much of the lake bed carried downstream. Erosion continued until, at least in one place, red Trujillo mudstones underlying the Ogallala Formation were exposed.

Once open to the Red River, several streams flowed across the basin, depositing sands, silt and clay, which you can see today in road cuts around town. At least two terraces have been cut into the banks of the basin. The first, is about 40 feet above the floodplain of Tierra Blanca Creek and you climb on to it just beyond the Interstate underpass. The second, 60 feet above the floodplain, is a half-mile on towards the Park. The terraces may have been benches, developed while the lake was still in existence, or perhaps they developed as the creek bed deepened.

Left *This section of a United States Geological Survey topographical map displays the route from the town of Canyon into the State Park.*

The park entrance is about 12 miles from the Plains-Panhandle Historical Museum in Canyon.

Park Road 5 from the park entrance to the turnaround is 7½ miles. You can loop back 2½ miles on Alternate Park Road 5 from the turnaround.

PALO DURO VISTAS

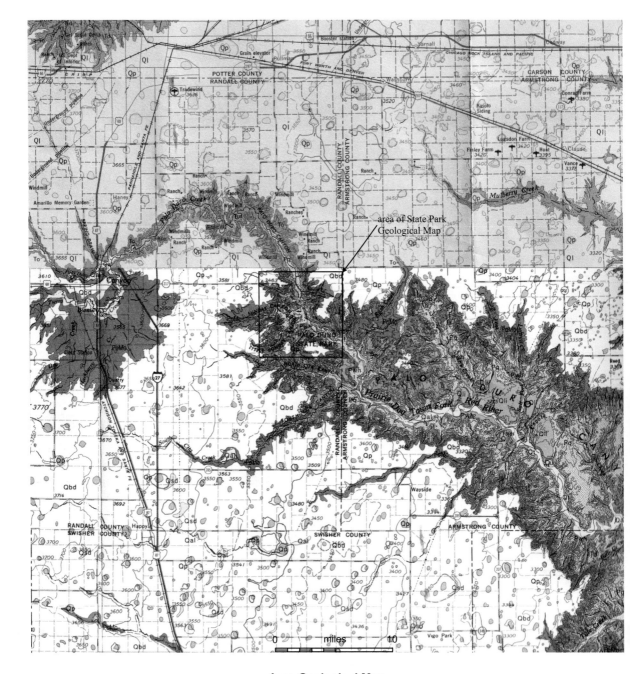

Area Geological Map

This map is compiled from parts of two Bureau of Economic Geology maps, the 1:250,000 Amarillo map to the north and the 1:250,000 Plainview map to the south, reduced somewhat. The actual scale of this map is about 1:320,000. The color used for the Blackwater Draw Formation changes from map to map.

Note the enormous number of playas.

Period/Epoch	Formation	Age m.y.	Description
Holocene	Lingos	.01-0.0	Alluvial deposits, stream-deposited gravel at base with wind-blown sand and silt at top; includes lake deposits and buried soils. Not present in Palo Duro State Park.
Pleistocene	Playa deposits	1.8-.01	Clay and silt, gray in shallow depressions, usually covered by a thin layer of recent sediment; weathers light gray.
	Blackwater Draw	1.8-.01	Sand, fine- to medium-grained quartz, silty, calcareous, includes caliche nodules, unbedded, pink to grayish-red, reddish-brown, olive-gray, distinct soil profile; thickness 25 ft., feathers out locally.
Pliocene to Miocene	Ogallala	5-11	Sand, silt, clay, gravel, and caliche. Sand, fine- to coarse-grained quartz, silty in part, cemented locally by calcite and by silica, various shades of gray and red. Minor silt and clay with caliche nodules, massive, white, gray, olive-green, and maroon. Gravel, not everywhere present, composed of pebbles and cobbles of quartz, quartzite, minor chert, igneous rock, metamorphic rock, limestone, and stream-worn *Gryphae* (a fossil oyster of Cretaceous age probably from Cretaceous rocks to the west) in channel deposits and in basal conglomerate; thickness 75-350 ft.
Triassic	Dockum Group	216-224	Sandstone, mudstone, shale, and conglomerate. Sandstone, fine- to coarse-grained quartz, micaceous, silty, thin-bedded to unbedded, indurated, gray, greenish-gray, brownish-red; mudstone and shale, sandy, silty, micaceous, locally calcereous, indistinctly bedded to unbedded, various shades of red, reddish-brown, orange, green, gray, yellow, purple; conglomerate of white, black, red and yellow chert pebbles and white quartz pebbles, sandy; basal conglomerate where present, also contains petrified wood and slabs of shale and sandstone; thickness 300 ft.
Permian	Quartermaster	254-251	Shale, siltstone, sandstone, gypsum and dolomite. Shale and siltstone, sandy, locally micaceous, evenly bedded, red, reddish-orange; shale indurated, includes thin beds and veins of satinspar; sandstone, fine-grained quartz, scattered frosted and polished grains; gypsum and dolomite beds thin and discontinuous; thickness 300+ ft.
	Cloud Chief		Sandstone, shale, gypsum and dolomite. Sandstone, fine-grained, silty, orange, orange-brown, various shades of red-mottled grayish green; shale sandy in part, red and orange-red; gypsum white gray and pink, up to 20 ft. thick; dolomite, rare, granular, discontinuous beds.

m.y. = millions of years

PALO DURO VISTAS

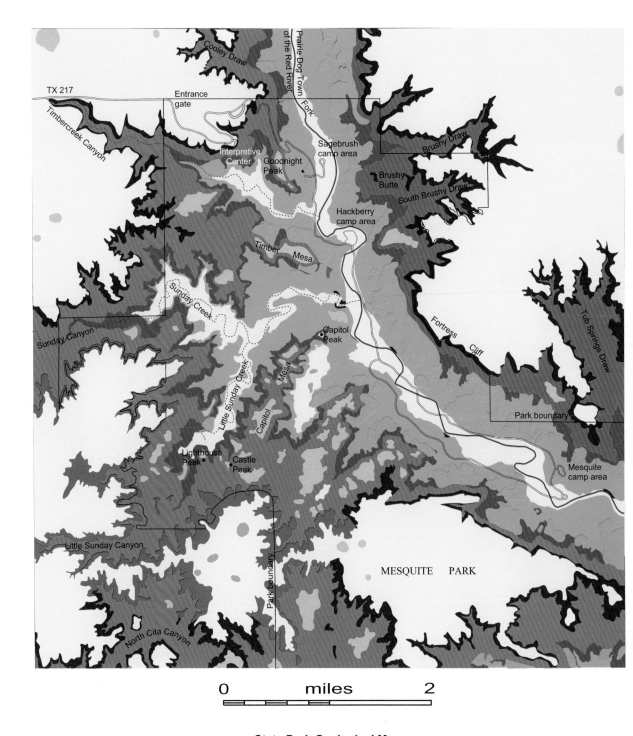

State Park Geological Map

Based on the Bureau of Economic Geologic map of Fortress Cliff Quadrangle, Scale 1:24,000, somewhat reduced. The actual scale of this map is about 1:37,500.

Period/Epoch	Formation	Age m.y.	Description
Holocene	Alluvium	.01-0.0	Sand, silt, gravel and clay along drainage ways. Includes some low level terrace deposits.
	Playa deposits		Clay, silt and fine sand.
	Landslide etc. debris and blocks		Landslide, slump and slope-wash debris and blocks, mostly of Tecovas and Trujillo strata with lesser amounts of Quartermaster and Ogallala rocks that have slid as large rotated blocks and as smaller debris deposits.
Holocene/Pleistocene	High Alluvium	1.8-0.0	Local deposits of sand, silt and some pebbles; up to 5 ft. thick.
Pleistocene	Blackwater Draw	1.8-.01	Freshwater lake deposits of silt, limestone and loess, underlain by caliche.
Pleistocene to Pliocene	Cita Canyon lake beds	5.3-.01	Lake beds of silt, clay and sand of Pliocene to early Pleistocene age overlain by Pleistocene lake deposits with interbedded wind-blown sand, silt and mud. The latter contains Perlette ash (a volcanic ash).
Pliocene to Miocene	Ogallala	5-11	Sand, silt, gravel, mud and caliche with thin gray shale lenses; cemented in places by silica; pockets of chert at base.
Triassic	Trujillo	216-220	Sandstone, fine- to coarse-grained, thick bedded, gray, stained red or dark brown on joint surfaces by iron oxides, with interbeds of dark maroon shale and marl-pebble conglomerate; sandstone layers form conspicuous cliffs; <200 ft. in total.
	Tecovas	220-224	Consists of, at base, shale, lavender, gray and white, with thin interbeds of soft sandstone, followed by a prominent white jointed sandstone, 25 ft. thick, and an upper section of orange shale; up to 200 ft. thick.
Permian	Quartermaster	254-251	Shales, brick-red to vermilion, with interbedded lenses of gray shales, clays, mudstones and sandstones; white veins of the satin-spar variety of gypsum criss-cross the formation; about 60 ft. thick.
	Cloud Chief		Sandstone, siltstone, mudstone and gypsum beds; gypsum veins common; minor folds from subsurface evaporite dissolution and associated subsidence of strata; some small-scale folding of gypsum beds due to conversion of anhydrite to gypsum; only upper 14 ft. exposed.

m.y. = millions of years

Playas on the Llano Estacado

Once you reach the High Plains surface at the top of the rise, not much is to be seen until the road bisects a playa about 2½ miles beyond Farm Road 1541. This playa also shows up on the aerial photograph on page 24, just to the right of the label for FM 217.

Playas are an interesting feature of the Llano Estacado. Over 20,000 have been mapped, 561 in Randall County alone; a rough count west of the canyon found more than one per square mile. Generally round and small, they are shallow depressions, intermittently water-filled, that range in size from 0.30 to 843.4 acres, averaging about 19 acres.

Contributing roughly 85 or 90 per cent of the recharge of the underlying Ogallala aquifer in the Southern High Plains, playas have clay-lined basins and periodically fill with water from rainfall and its associated runoff. As water accumulates in playas during rainy periods, it infiltrates the clay floor through cracks, plant root openings and other pores and flows through fissures in the Caprock to the Ogallala beds below. Eventually, cracks in the floor swell shut as the clay itself absorbs water, reducing the recharge. Once a playa is full, recharge occurs along its perimeter where there is little or no clay. In some places where Blackwater Draw material is thin or absent, basins have developed in the Caprock.

There are many theories about the origin of playa lakes, the most popular being that they result from wind action on soils disturbed by drinking animals. In this hypothesis, rainfall runoff gathering in a

Left *The southern section of the playa bisected by FM 217. This shallow playa is probably quite young.*

low spot on the plain attracts animals such as buffalo which break up the soil surface with their hooves. When the water dries up, the broken surface is blown away by the prevalent Panhandle wind. Such saucer-shaped depressions, called blowouts, are quite common on the Great Plains. Over time, blowouts expand to form playas.

The initial low spots may, in some cases, have been sinkholes, created by subsidence underground. Sinkholes are created on the High Plains by water that contains carbonic acid dissolving the underlying caliche. The surface subsides, creating circular depressions that expand with time. The carbonic acid comes from the oxidation of organic material.

Whatever their origin, playa lakes are important because they store water in a part of the country that receives as little as twenty inches of rain a year and has no permanent rivers or streams. They support an astounding array of wildlife. Two million waterfowl winter there and you can find mayflies, dragonflies, salamanders, bald eagles, endangered whooping cranes, jackrabbits and raccoons at playa lakes. The eastern Panhandle is on the Central Flyway, one of the routes by which birds migrate, so twice a year, migrating birds use the playa lakes for water.

Timbercreek Canyon

Timbercreek Canyon comes into view as the highway dips down to Sunday Canyon Road. The edge of the canyon is very close to the highway so this is a good place to discuss how canyons form and the rate at which they advance.

The walls of these canyons advance through two main processes. The first involves spring sapping, seepage erosion and piping. In spring sapping and seepage erosion, water seeping out at the base of sand beds in the Ogallala, and sandstone beds in the Trujillo and Tecovas Formations carries with it particles from the beds and erodes them. Also, because seepage areas are wet, plants grow there; their roots contribute to the erosion. In piping, percolating water creates narrow tunnels or pipes by physical removal of grains and soluble material. Pipes exit through the escarpment, sometimes creating large caves, one of which is shown on page 69.

The second main process involves the taking into solution of underground evaporites, discussed earlier (page 11). As the salts thin, the surface above subsides, causing the Caprock to fracture. Eventually, a body of rock breaks off from the wall of a canyon and slumps down the escarpment. Palo Duro Canyon has many such slump blocks, as the Park geological map shows, although none have been mapped in Timbercreek Canyon.

The spectacular gash of Palo Duro Canyon dominates the view from the air in this photographic mosaic. The principal subsidiary canyons that branch off the main canyon to the west are labeled, as is the town of Canyon.

Palo Duro Canyon has two straight sections, one along Palo Duro Creek, the other north of Timbercreek Canyon. Perhaps the drainage follows fault zones along these sections. Both show up prominently on the shaded relief map on page 12.

A critical question for residents of the Panhandle is the speed at which the Caprock Escarpment is retreating. As a rough estimate, the Ogallala Formation has been eroded back about 40 miles from its original eastern boundary in the last 4.5 million years (see the Ogallala map on page 40), i.e. slightly more than a half-inch per year.

Actual measurements over the last 15 years found that it is retreating at between 0.38 and 1.20 inches per year, roughly the same rate.

The State Park Entrance

The State Park entrance is about 12 miles from Canyon. A national park in the upper Palo Duro Canyon was the dream of local citizens early in the settlement of the area. However, the Federal government was reluctant to buy park land from private individuals, rather than convert existing federal lands to parks as in other states.

Local enthusiasts then approached their state government representatives with the idea of creating a state park. Texas did not have any state parks until 1923, when a State Park Board was established to accept donations of land for parks. In 1933, the Park

Above *The head of Timbercreek Canyon from Sunday Canyon Road. The upper level in the wall of the canyon at mid-left is of Blackwater Draw material up to 10 feet thick, eroded in places. The 75 feet decline approaching the canyon is probably in this formation.*

The Blackwater Draw strata lies above an even, brown bed some 10 feet thick, the Caprock caliche, also seen on the other side of the canyon at right. It is normally a light gray or white color as seen at the bottom left of the photograph but has been stained by runoff from the overlying sediments.

Below the Caprock are brown sands and silts of the Ogallala Formation.

Notice how narrow and deep the canyon is, only 200 feet wide and 95 feet deep as it turns the corner in mid photograph, and how abruptly it ends at road side.

Board was persuaded to buy 16,402 acres of land from a landowner for $377,000, having first secured agreement from the Civilian Conservation Corps that four companies of 200 men would be assigned to build amenities in the Park. The Civilian Conservation Corps was a federal New Deal entity formed to alleviate unemployment during the Great Depression, especially among First World War veterans.

The Park Board agreed to service the debt by assigning 50 per cent of gross entrance revenue and 20 per cent of all concession income for that purpose. The bulk of the debt was held by the landowner, the remainder by two Scottish mortgage companies.

The first project completed was the building of a road to the bottom of the canyon, followed by the construction of the Interpretive Center. Both were finished for the opening day, the Fourth of July, 1934, to much local jubilation. By 1937, the CCC had also completed trails, the four Cow Camp cabins, the Park headquarters, the entrance building and shelters.

Below *The Park entrance building was constructed by the Civilian Conservation Corps in 1935-37.*

Above *This photograph, taken while the road down into the Canyon was being constructed, shows how many men worked on the project.*

Photograph courtesy of the Panhandle-Plains Historical Museum.

Despite these improvements, however, the Park did not generate enough revenue to service the debt. Unpaid interest was added to the principal, so that by 1945, total debt was about $580,000. After much negotiating, however, the landowner agreed in 1945 to sell the entire property to the Texas State Park Board for $300,000.

The purchase was funded by bonds that were serviced by 90 per cent of entrance fees and 20 per cent of concession income. This time, thanks to post-war prosperity and to inspired promotion by the Park concessions operator, the Park flourished, and the bonds were retired well ahead of schedule in 1960 and 1966.

The idea to build an amphitheater to stage plays came to local residents in 1960 and the next several years were devoted to raising funds to do so. Full summer programs, 5 days a week from early June to mid-August, began in 1965.

The 2,036-acre Cañoncita Ranch along the Park's southern boundary was added to the Park in 2002, funded by a grant from the Amarillo Area Foundation. The foundation funded the purchase of an additional 7,837 acres in 2005, including the site of the last Comanche battle in 1874. Palo Duro State Park now has a total area of 26,275 acres, making it the second largest operating state park.

The Interpretive Center

The Interpretive Center was originally called El Coronado Lodge. It was built by the Conservation Construction Corps in the period 1933-34 to designs by a prominent Amarillo architect and park booster, Guy Carlander. The center was later expanded to include interpretive exhibits and murals.

The three guest cabins to the west of the Interpretive Center were built by the Civilian Conservation Corps in 1936-37. Perched on top of the cliff, the cabins provide magnificent views over the canyon. All the buildings constructed by the Civilian Conservation Corps were made of sandstone quarried in the Park.

Right *Brown Ogallala sandstones and siltstones photographed on the embankment behind the Interpretive Center.*

Interpretive Center Viewpoint

The magnificent vista over the lower canyon is seen in this photograph from the Interpretive Center Viewpoint. The map on the following page shows the geology from the same orientation.

The ledge to the front left, and Goodnight Peak beyond it, are capped by light gray sandstone and reddish brown shale of the Trujillo Formation, overlain by patches of High Alluvium sand and silt.

Behind Goodnight Peak on the horizon, Fortress Cliff has a conspicuous white cap of light-colored sand and siltstone of the Blackwater Draw Formation underlain by the Caprock.

Timber Mesa, in the right middle distance across Timbercreek Canyon, has a similar sequence of strata to Goodnight Peak. At its base, red Quartermaster shales are overlain by Tecovas shales. The erosion of the Quartermaster and Tecovas rocks has created structures that remind visitors of the flared skirts worn by traditional Spanish dancers, and are called the Spanish Skirts. One can be seen at the right of the photograph.

Timbercreek Canyon runs between Timber Mesa and Goodnight Peak to join the main canyon. On the skyline behind Timber Mesa is the flat-topped Mesquite Park mesa.

Left *The geological map of the State Park has been turned around to give you the same line of sight as you get from the parking lot above the Interpretive Center, shown at the bottom right of the map.*

Notice how the Prairie Dog Town stream takes a wide detour around the nose of Timber Mesa at Hackberry

View from the Interpretive Center Viewpoint

Do try to spend some time here to view the magnificent vista over the canyon shown in the photograph on the two previous pages. The map above shows the geology with the same orientation.

The Quartermaster Formation, the oldest formation on view here, is about the same age as the youngest formation in the Grand Canyon. Palo Duro Canyon is often called in Texas "the Grand Canyon of Texas". It certainly is the largest canyon in the interior of Texas, although the canyons along the Rio Grande are much longer and several are deeper. The viewpoint is about 540 feet above the junction of Timber Creek and Prairie Dog Town Creek. As you go downstream, the canyon deepens. At the turnaround seven miles ahead, the cliffs on the Mesquite Park mesa are 660 feet above Prairie Dog Town Creek.

Right *This map shows the various soil types found in the Central and Southern High Plains, the latter being south of the Canadian River.*

Loam is a term used by soil scientists to describe a soil made up of equal parts of clay, silt and sand.

The soils become sandier as you approach the Pecos River valley, indicating that most of the material making up the Blackwater Draw Formation came from this valley.

Blackwater Draw Formation

The descent into the main canyon begins about 0.9 mile from the Interpretive Center viewpoint. The Blackwater Draw, Ogallala, and Trujillo Formations, and part of the Tecovas Formation, are well exposed during the descent, and this is your opportunity to see them up close. Once down in the canyon, only the Permian Quartermaster Formation is at road level.

First on view as you turn to the right is the sand and silt of the Blackwater Draw Formation, an accumulation of wind-blown sand and silt that covers most of the Southern High Plains. The formation can be as much as 75 feet thick in places but is more commonly 30 feet thick or less and includes as many as eleven buried soils.

During the Ice Ages, glaciers advanced and retreated numerous times, at their maximum reaching as far south as Kansas on the Great Plains. When they advanced, the climate in the Panhandle became cooler and wetter in summer than now, and warmer in winter. Grasslands developed, wind erosion was minimized, and calcium-rich soils developed at the surface.

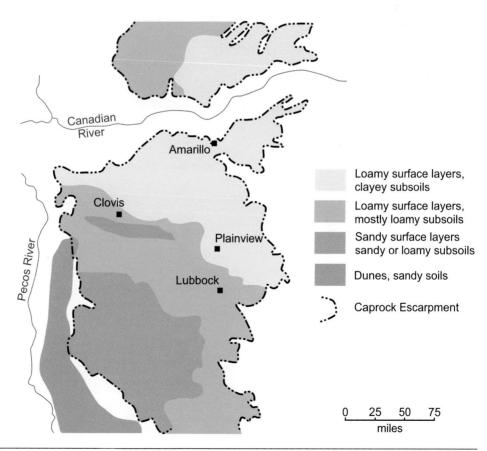

When glaciers retreated, the climate became warmer, drier and windier, and grasslands diminished. Erosion in the broken ground of the Canadian and Pecos valleys produced dust that accumulated as Blackwater Draw sediments on the Llano Estacado.

The formation also includes two volcanic ash layers, the Lava Creek Ash which erupted from the Yellowstone volcanic center about 640,000 years ago, and the Guaje volcanic ash which came from the Jemez Mountains in northern New Mexico 1.62 million years ago. The formation is therefore at least 1.62 million years old at its base.

Fossils in the Blackwater Draw Formation

An astonishing assemblage of exotic animals inhabited North America during early Blackwater Draw times, many of them now extinct. Included in this category were mammoths and mastodons that towered elephant-like over almost all else on prairies and in

Left *This photograph of the Blackwater Draw Formation was taken before beginning the descent into Palo Duro Canyon, about a mile east of the Visitor Center.*

It shows the typical unconsolidated, loose, fine silt of the formation in a road bank.

boggy woodlands; several types of slow-moving giant ground sloths as large as mammoths; 2,000-pound armored six-foot-long glyptodonts resembling nothing known today; six-foot-long capybaras; 500-pound tapirs; 300-pound giant beavers; dire wolves whose large heads and powerful jaws made them resemble hyenas; 1,500-pound short-faced bears; scimitar-toothed cats that fed on mammoth young; and great saber-toothed cats that could rip open their prey with their enormous teeth.

Many of these animals developed in Asia and migrated to the Americas during the Pleistocene epoch, the time of the Ice Ages, which lasted from 1.8 million years to 10,000 years ago. During periods when glaciers advanced, the ocean levels were lowered as water became tied up in ice. Land bridges, some a thousand miles across, emerged from the ocean and allowed animals to cross from Asia to North America.

The Blackwater Draw sands and gravels around Palo Duro are rich in these fossils, most prominently mammoths and bison. Mammoths belong to the elephant group of animals. There have been more than 500 different kinds of elephants on earth at various times over the last 55 million years, of which two have survived, the Asian and African elephants. Mammoths developed from the Asian elephant and were comparable in size to the present-day African Savannah Elephant, which weighs 4 to 7 tons and is 10 to 13 feet tall.

The first mammoth in North America was the Southern Mammoth (*Mammuthus meridionalis*), which arrived about 1.8 million years ago. The second was the Steppe Mammoth which may have developed from the Southern Mammoth or migrated to North

Right *One of the largest elephants ever to have lived, the Columbian mammoth was like a large African elephant but with a more sloping back and long, spiraled tusks. One tusk from a mammoth found in Texas measured 16 feet, the longest of any of the elephant family.*

There is some debate as to how much hair the Columbian mammoths had. Some scientists suggest that they had a fur coat like the woolly mammoth but is more likely that hair grew on some parts of their bodies, such as the tops of their heads, but that they were basically elephant-like with exposed naked skin, grayish in color.

Left *The American mastodon, Mammut americanum, was wide-spread across North America from Alaska to central Mexico. Mastodons were smaller than mammoths, reaching about ten feet at the shoulder, and their tusks were straighter and more parallel. They were about the size of modern elephants, though their bodies were somewhat longer and their legs shorter. Mastodons were herbivores; their teeth were used for clipping and crushing twigs, leaves and other parts of shrubs and trees. Most of the plants they ate were ones that grew near swamps and wet areas in woodlands.*

America about 1.2 million years ago, the most advanced of which, *Mammuthus jeffersonii*, survived on the plains until about 11,000 years ago.

Mammoths were grazers, feeding mainly on grass and flowering plants. During a 60- to 80-year life span, a mammoth would go through up to six sets of flat molar teeth. Adult teeth were about the size of a large shoe and had a flat, ridged surface adapted for grinding up tough plant material. The largest species of mammoth (*Mammuthus imperator*) reached 13 feet in height; the smallest pygmy mammoths (*Mammuthus exilis*) grew 4 to 8 feet in height. Mammoths had long, curved tusks used for protection, to establish dominance, and to help gather food.

No mastodon fossils have been found in the State Park but Harley Burton, in his history of the JA Ranch states that in the old ranch headquarters building "On a shelf extending entirely around the wall of this room is a large collection of mastodon bones and rocks found four miles south of headquarters." The Panhandle-Plains Historical Museum in Canyon has an exhibit of a shovel-tusked mastodon.

The bison, whose ancestors are thought to have originated in southern Asia some 400,000 years ago, arrived in North America during the Kansan glacial age, which lasted from 410,000 to 380,000 years before the present. The ancient bison was much larger than the present-day animal and ranged throughout the northern hemisphere. One prehistoric bison, *Bison latifrons*, had horns measuring 9 feet from tip to tip.

Right *The modern bison is a ferocious animal; full-grown bulls weigh up to 2,000 pounds and stand six feet or more at the shoulder. Their massive heads, with their thick covering of matted wiry hair, hold a set of horns that are never shed. The shoulders carry a huge hump that gives the bison a top-heavy look, the hips being much smaller than the shoulders.*

A bison cow is smaller, weighing up to 1,100 pounds and is 4 - 4½ feet tall at the shoulder. Cows usually conceive for the first time as three-year olds. Though calves can be born at any time of the year, the calving season usually begins in mid-April after 9 - 9½ month gestation period.

Calves are able to walk minutes after they are born and remain with their mothers for about a year, or until another calf is born. Bison are herbivores or plant eaters, and feed primarily on wheat and buffalo grass, blue grama, and other similar grasses. Though they generally have poor eyesight, bison have excellent hearing and a keen sense of smell. They reach maturity at seven or eight years and may live to the ripe old age of thirty.

A more modern bison, *Bison occidentalis*, evolved in the late Pleistocene and was the immediate ancestor of our present-day bison. In 1970, a *Bison occidentalis* skull with a horn spread of 30½ inches was found in a stream bank of the upper Palo Duro Canyon just north of the State Park. Along with the skull was found the remains of small fish and 23 species of freshwater and land gastropods, a class of animal that includes snails, slug, limpets and other seashell animals. The gastropods were found through radiocarbon dating to be about 10,800 years old.

The type of gastropod present shows that the climate here was cooler and wetter than now and that the area was marshy with a good deal of standing water. The land species present in the fauna occupied habitats ranging from moist, deep woodlands to sparse, well-drained woodlands.

Fossils of two pre-Columbian horses, Scott's horse, *Equus scotti*, and the Mexican horse, *Equus conversidens*, have been found in Randall County. Scott's horse was about the same size as modern horses, perhaps even a little larger. It is believed to have been descended from European or Asian horses that crossed over from Asia in the early Ice Ages. A specimen has been found in the Rio Grande valley in sediments that are between 1.6 and 1 million years old. The animal flourished in well-grassed North America although the horse began as a forest animal related to the rhinoceros and tapir groups of animals. *Equus conversidens* ranged from Canada to Mexico. Remains have been found in a Canadian campsite, 11,300 years old.

Vast numbers of fossils of other smaller animals have also been recovered from the Blackwater Draw Formation. In Randall County alone, mice, shrews, voles, squirrels, prairie dogs, gophers, muskrats, badgers and peccaries have been found.

Most of the *megafauna*, as the large animals are called, disappeared around the end of the Pleistocene, which also marked the end of the Ice Ages, and all are now extinct. The extinction roughly coincides with a dramatic increase in northern hemisphere temperatures around 11,700 years ago, as recorded by Greenland ice cores. The climate became warmer by 17°F in only 250 years and by another 25°F in the following 2,500 years. The extinction was not restricted to megafauna. Many birds, including jays, ducks, flamingos and raptors, also disappeared.

Many theories have been advanced to account for the disappearance of the animals, including extermination by humans. But as this book is being written, a new and controversial theory has been advanced that a comet exploded over the Great Lakes destroying the habitat. The evidence is that many of the fossil sites are overlain by a black charcoal layer, containing exotic materials such as tiny spheres of carbon containing diamonds less than a micron in diameter.

Right *Overlain by about 18 inches of the Caprock, lightly-cemented brown sand and silt of the Ogallala Formation crop out beside the road.*

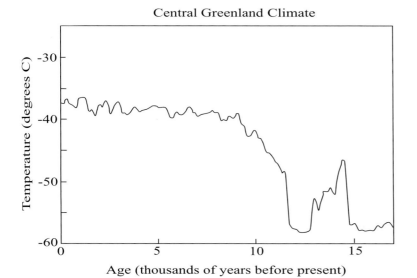

Left *Dramatic climate fluctuations, recorded by snowfall in central Greenland, occurred in the last 5,000 years of the Pleistocene epoch.*

The sudden rise in temperature between 11,700 and 9,000 years ago coincided with the disappearance of many animals in North America.

The Ogallala Formation

Brown sandstones of the Ogallala Formation are well exposed in road cuts and banks beside the Interpretive Center (photograph above), and below the Blackwater Draw sands and silts on the descent into the canyon about a mile beyond the Interpretive Center Viewpoint (photograph on page 42). The formation was created between about 17 and 4.5 million years ago in one of the most extraordinary events in geologic history. (From this point in the book, I will use the abbreviation Ma for "millions of years ago").

Beds of the Ogallala Formation lie at a slight angle to the underlying red and green shales of the upper Trujillo Formation, which are 200 million years older. Much happened in Texas during this enormous span of time, although no traces remain in the canyon. During the Cretaceous period, the middle of the continent subsided and the ocean flooded in, creating a continental sea called the Western Interior Seaway stretching from the Gulf of Mexico to the Arctic

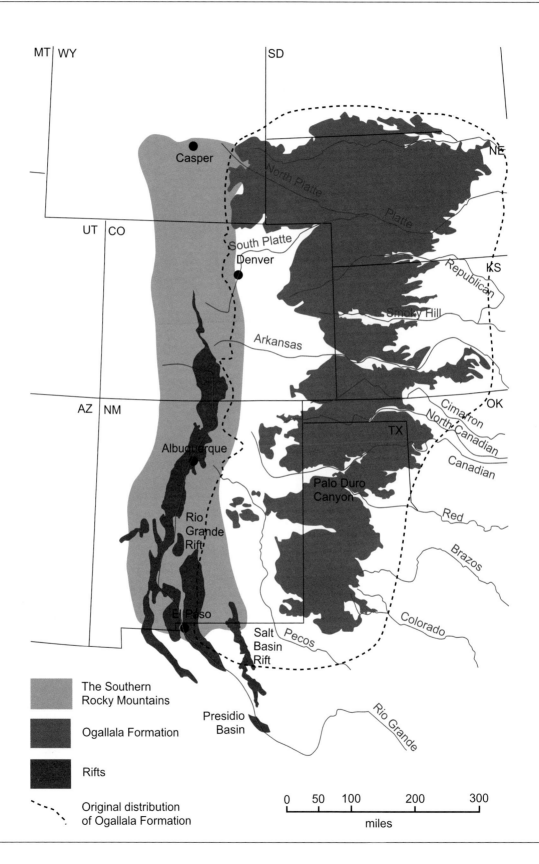

Ocean. The limestones and sandstones you see throughout central Texas were created in this sea. The ocean withdrew around 80 Ma when the Earth's crust in central North America uplifted.

The Rocky Mountains arose between 50 and 40 Ma as a series of fault blocks produced by compression across North America. A period of erosion followed in which the Rockies were reduced in height and much of the Cretaceous strata and some of the Permian strata on the Southern High Plains were eroded away, creating broad valleys and uplands.

The Rio Grande Rift, named after the river that flows down it in Colorado and New Mexico, began developing about 27 Ma along the crest of the southern Rocky Mountains. The rift, a long, narrow trough in which the Earth's upper crust was extended or stretched, began as a series of broad, shallow basins along its length until about 15 Ma when the Southern Rockies began to uplift. The rift deepened as the mountains arose and it now consists of four major linked basins up to 27,000 feet deep, filled with lake and river sediments and more or less inactive.

Streams flowing off the uplifting Rockies carried sand and gravel to the east, filling up valleys and creating the Ogallala Formation. Valley deposits can be as much as 600 feet thick, consisting of sand and gravel deposited in stream beds and sand and clay dumped where streams flooded their banks. Interbedded with these sediments are wind-borne or eolian sand and silt sheets. Uplands between the valleys are mostly covered by eolian deposits up to 100 feet thick. The formation contains numerous buried soils which indicate that the region was grassland with a semi-arid to sub-humid climate.

The Rocky Mountains continued to rise after the end of Ogallala deposition ended at 4.5 Ma and brought up the Ogallala and the entire Great Plains with them. Studies of the Ogallala in Nebraska and Wyoming show that the formation has been tilted as much as 2,250 feet up to the west since its deposition. In fact, the Ogallala could not have attained its great width without this tilting; the sediments were not thick enough to be spread across 400 miles by water power alone. The uplift of the continent extends as far as Dallas to the east and California to the west, centered on the Rio Grande Rift.

How did all this happen, you may ask, and are these events interconnected? Yes, they are interconnected; they originated about 180 Ma when North America began over-riding an oceanic plate to the west, the Farallon Plate. The continent came under compression as a result which led to folding and thrusting and the creation of the Rockies at 50-40 Ma. At 27 Ma, North America reached the western margin of the Farallon Plate and subduction ended.

Left *The Southern Rocky Mountains, created between 50 and 40 Ma, were eroded continuously until about 15 Ma, when they began being uplifted again. The Rio Grande Rift, a series of basins now as deep as 27,000 feet, began forming along the crest of the mountains about 27 Ma in New Mexico and developed northwards into Colorado. The subsidiary and much less developed Salt Basin Rift formed to its east at roughly the same time.*

Erosion of the uplifting Rockies provided sand and gravel that was carried by eastward-flowing rivers to create the Ogallala Formation from about 17 to 4.5 Ma. Best known in Texas as the main water source for the High Plains, the Ogallala is largely intact and forms one of the most widespread alluvial bodies on Earth.

The map shows the enormous extent of its present-day outcrop area, 800 miles from north to south and 400 miles across at its widest in Wyoming and Nebraska. It also shows the estimated original extent of the formation, which is not greatly larger than today's area.

Right *The Caprock is an accumulation of calcrete (calcium carbonate - caliche in Spanish) that formed above the Ogallala Formation during a long period of climate stability.*

The plate broke away from the surface and now lies under the east coast of North America. In its wake, hot material from the interior of the Earth rose up towards the surface and because it was hotter than the crust it buoyed up the area above it. Both the Rio Grande Rift and the general uplift of the Great Plains resulted from this buoyancy.

The Ogallala Formation on the Llano Estacado is about 11 million years old at base, according to fossil evidence, and continued being laid down until around 4.5 Ma. It began accumulating earlier in the north, at 17 Ma in Nebraska, for example.

In Palo Duro Canyon, the Ogallala is up to 90 feet thick, and consists mainly of medium- and coarse-grained sand with beds of loess. The formation is capped by about 6 feet of calcrete or caliche, a calcium carbonate deposit. Calcium is brought in by wind-blown dust and rainfall. The calcrete forms through percolating groundwater containing dissolved calcium bicarbonate coming to the surface and precipitating calcium carbonate. The carbonate cements together other materials, including gravel, sand, clay, and silt and over time accumulates as a layer of hard material, known to geologists as *hardpan*. Hardpan occurs worldwide, generally in arid or semi-arid regions.

The Caprock formed over several hundred thousand years between the end of Ogallala deposition, at about 4.5 Ma, and the beginning of Blackwater Draw deposition, somewhere around 1.8 Ma. This was a period of landscape stability in which not much wind-blown dust was brought on to the Llano. It was probably a period of moderate rainfall brought on by cooler temperatures during the early Ice Ages.

Left *Photographed from Park Road 5 as it begins its descent into Palo Duro Canyon, the Caprock, about 18 inches thick, crops out below light gray Blackwater Draw strata, and overlies light brown sands and silts of the Ogallala Formation. Note the openings in the lower Ogallala strata, the result of piping.*

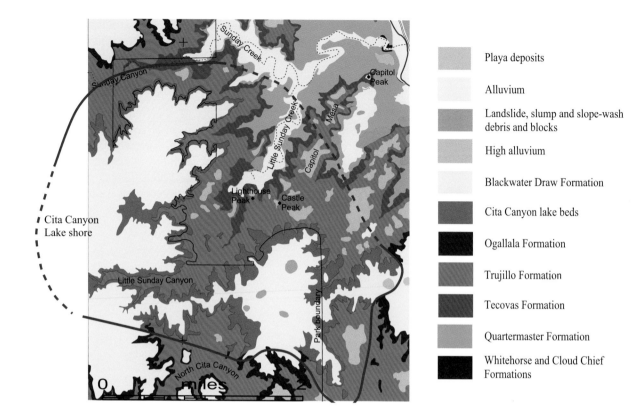

The Cita Canyon Lake Beds

In the southwest part of the State Park, around Lighthouse Peak, the Ogallala Formation is absent and in its place is a set of mud, silt and sand beds called by geologists the Cita Canyon Lake Beds. Such lake sediments, quite common on the Llano Estacado, were deposited in shallow basins that probably formed as a result of salt in the Permian rocks below being dissolved by groundwater. The sediments show evidence of flooding, and because all formed at about the same time over a wide area, they indicate that the water table rose at the end of the Ogallala deposition, either because of cooler temperatures or increased rainfall or both. Blocks of the Caprock are found in the lake beds suggesting that the beds are slightly older or of the same age as the Caprock.

The main interest in the Cita Canyon beds is in the extraordinary variety of fossils found there. The basin itself was nearly 4 miles across, set in a grassy plain and intermittently full of water. Animals that lived on the plains came here to drink. Sometimes they died or were killed at the lake, their skeletons buried in the mud and fossilized over time, and later uncovered by erosion.

Right *Exhibit in the Panhandle-Plains Historical Museum in Canyon of a ground sloth skeleton found in Cita Canyon.*

The ground sloth was an enormous creature, up to 10 feet long, whose ancestors had migrated up from South America about 8 Ma and is now extinct.

Photograph courtesy of the Panhandle-Plains Historical Museum.

Left *The roughly circular outline of the Cita Canyon lake is superimposed on a section of the State Park geologic map.*

The lake beds of mud, silt and sand, 4.8 million years old at their base, have intervals of wind-blown materials, and near the top a volcanic ash bed about 640,000 years old.

This treasure trove of fossils discovered in the 1930s led to a project in which up to 75 men were employed year-round by the Works Progress Administration, a New Deal agency set up to alleviate unemployment during the Great Depression. Between 1936 and 1940 under the supervision of C. Stuart Johnson, the curator of Paleontology and Archeology at the Panhandle-Plains Historical Museum, the project produced fossil remains of horses, camels, mastodons, deer, antelope, peccaries, ground sloths, glyptodons, tortoises and water birds.

The ground sloth was a massive creature, up to 10 feet long, whose ancestors had migrated up from South America about 8 million years ago and is now extinct. It may have lived on in Cuba until 1550.

The glyptodon, a relative of the armadillo but up to 5 feet long, was an herbivore, grazing on grasses and other plants found near rivers and small bodies of water. A large and heavy mammal, the glyptodon probably could only have moved at one or two miles per hour.

These fossils are believed to have belonged to the period 4.75 to 1.8 Ma, i.e. just prior to the Ice Ages, which began at 1.8 Ma and so are older than the Blackwater Draw fossils described earlier.

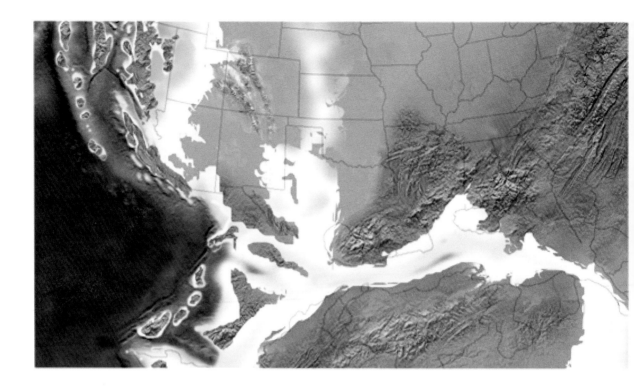

The Older Rocks in the Park

Just as the road into the canyon makes a hard turn left at the base of the Ogallala Formation, upper Trujillo sandstone cliffs can be seen on the right (photograph on pages 52-3). These rocks are a 200 million-year step back in time to an entirely different environment, one where almost all today's continents were part of a giant continent called Pangea, and where south Texas straddled the equator.

From late in the Carboniferous period through the beginning of the Permian, the eastern and southern margins of North America were over-ridden by a continental plate that approached from the south, probably part of what is now South America and Africa. The collision caused great stress in North America. Mountains formed along the collision zone, first the Appalachians and then the Ouachita-Marathon Mountains. The latter, which date to about 285 Ma, have been long since eroded away in Texas but their trace, called the Ouachita Front, is well known from oil well intersections and exposures at Marathon in west Texas.

At the time of the collision, the Earth's crust buckled north of the mountains, creating uplifts and sub-basins in the sea, such as the ones shown in the map on page 48. The sub-basins began subsiding rapidly during the late Carboniferous period and filled with sandstones and conglomerates eroded off the adjoining uplifts.

Left *This diagram of the Earth in the late Permian period (about 225 million years ago, when the oldest rocks in Palo Duro Canyon were created) shows most of the continents fused together in a giant body called Pangaea ("all lands" in Ancient Greek).*

The range of high mountains straddling the equator developed when South America and Africa came up against North America. Those in Texas are the Ouachita-Marathon Mountains, the ones to their right, the Appalachians.

Diagram by R.C. Blakey, website vishnu.glg.nau.edu/paleogeogwus.htm.

By the end of the early Permian, at about 280 Ma, the sub-basins had filled and shallow water extended across the area over level ocean floor. For the remainder of the Permian period, the area subsided slowly, and limestones, anhydrite ($CaSO_4$), halite (NaCl – common salt) and siltstones or mudstones were deposited in shallow ocean water.

Texas as it looked about 225 Ma during the late Permian period, i.e. 60 million years after the collision, is shown opposite. Land to the south was separated from Texas by a narrow ocean channel, and the Permian ocean occupied a strip of land up to the northern Panhandle. The Ouachita-Marathon Mountains ran from Texas to Louisiana and Arkansas and joined up with the Appalachians in Mississippi.

The Permian Basin, as the overall area of Permian deposition is called, is famous in Texas for containing vast amounts of oil and gas. The hydrocarbons, mostly originating from plankton in shales older than the Permian rocks, rose up and were trapped below the evaporites which form effective seals.

Most of the hydrocarbon production is found along the boundaries of the sub-basins, one of which, the Palo Duro Basin, underlies the State Park. Some hydrocarbons are being produced along the northern and southern borders of the Palo Duro Basin. The basin is relatively shallow in the Palo Duro area with about 3,600 feet of underlying Permian strata. Limestone deposition in the basin ended about 265 Ma and its upper 1,040 feet is filled with beds of shale, mudstone, siltstone and halite, and two thin anhydrite beds. The final retreat of the ocean across Texas at the end of the Permian period is marked by tidal and lagoonal deposits such as the shales and mudstones of the Quartermaster Formation and the underlying shales and gypsum beds of the Cloud Chief Formation (see pages 66-67).

Between the end of the Permian period at 251 Ma and the deposition of the lower Triassic Tecovas strata 27 million years later, one of the most important events in the history of life on the planet took place, the Permian-Triassic mass extinction. This mass extinction was one of five so far identified in the Earth's history, and was by far the most catastrophic.

It may have occurred in 2 phases, the first at 256 million years ago, and the second at the end of the Permian period, 5 million years later. The second phase happened quickly; Permian and Triassic fossils in terrestrial sedimentary rocks in Antarctica are separated by only 23 to 33 feet of rock, representing at most 200,000 years.

Two main hypotheses have been advanced to explain mass extinctions. One is that they were caused by collisions between the Earth and extra-terrestrial objects such as meteoroids or comets. Such collisions would have thrown up so much material into the

PALO DURO VISTAS

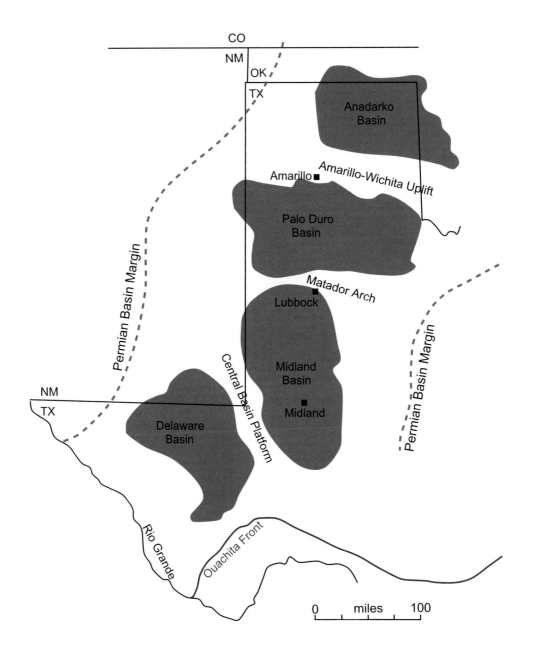

Permian Basin Sub-basins

The four main sub-basins in the Permian Basin in Texas are shown in this map. They are separated by platforms, such as the Amarillo-Wichita Uplift, that were under shallow water for much of the Permian period, but emerged as dry land towards the end of the period.

atmosphere that sunlight would have been diminished for some time. Material in the atmosphere would have disrupted photo-synthesis, the process by which plants and algae manufacture sugar from light, carbon dioxide and water, and produce oxygen as a by-product. Plants would have died as a result, eventually drastically compromising the entire food chain.

The end-Cretaceous extinction that wiped out the large dinosaurs was caused by such a collision. Walter Alvarez, the geologist who first developed this thesis, now believes that the object was most likely a comet; its crater has been found in the Mexican Yucatan. No such evidence has yet been found for the Permian-Triassic extinction, however.

The second main hypothesis put forward to explain mass extinctions is that they resulted from massive eruptions of basaltic volcanic rocks, flood basalts as they are called. Such eruptions would have produced huge amounts of carbon and sulfur dioxide, as well as aerosols that would have blocked a significant amount of sunlight, initially resulting in cooling. Although the sulfur dioxide and the particulate matter would have left the atmosphere in a few months, the carbon dioxide would have remained and would have caused global warming.

The Permian-Triassic boundary does coincide with the largest of the flood basalt eruptions, which occurred in Siberia at around 251 million years ago. There, about 950,000 cubic miles of mainly basalts erupted in a very short time, perhaps as short as a million years. If this volume of lava were to be spread evenly over the Earth's entire surface, it would produce a layer nearly 20 feet thick. This hypothesis looks promising but is by no means conclusive. One of the leading experts in the subject, Douglas Erwin of the Smithsonian Institute, wrote an entire book on the subject, published in 2006, in which he confessed that he did not know the cause of the extinction.

The Permian-Triassic extinction had an enormous impact on life. About 90 per cent of marine species and about 70 per cent of land animals disappeared. In the oceans, it took 20 million years before new kinds of reefs appeared, built by organisms different from those that built the Permian reefs. On land, no significant plant material was buried for the first 10 million years of the Triassic period. The forests having been devastated, there is a "coal gap" in the geological record, with no coal found in strata laid down in those years.

Permian reptiles, including the rhinoceros-sized plant-eaters as well as the sabre-toothed gorgonopsians that preyed on them, were gone. Only some small- to medium-sized amphibians and reptiles inhabited the Early Triassic world.

Thirty million years after the Permian-Triassic extinction, the environs around Texas at the time of the Tecovas Formation in the mid-Triassic are shown in the reconstruction opposite. Most of Texas, including the Panhandle, was dry land. The Ouachita-Marathon mountain belt crossed the middle of the state and continued up to the northeast as the Appalachians. To the south, a trough was beginning to develop, the Chihuahua Trough; it would deepen and fill with evaporites and limestones during the Jurassic and Cretaceous periods. Along the Appalachians to the northeast, rifting had begun that would eventually create the Atlantic Ocean, separating North America from Europe and Africa. The sediments making up the Tecovas Formation came from north and east-northeast of the Panhandle.

By the time of the upper Triassic Trujillo Formation (in the lower diagram opposite), the Gulf of Mexico opening had widened and the Ouachita-Marathon and Appalachian mountains had been uplifted and streams and rivers now flowed off them across the Panhandle, bringing sand and silt with them. In places, the Trujillo sandstones include pebble-sized metamorphic rock fragments and abundant flakes of mica, a silvery mineral created by metamorphism. The term metamorphic is used by geologists to describe rocks that have been altered from their original appearance and composition by heat and/or pressure. Metamorphic rocks are rarely found on the surface in Texas but are known to occur deep in the Ouachita-Marathon Mountains, leading geologists to conclude that the Trujillo mica came from deeply eroded parts of these mountains.

Both Tecovas and Trujillo strata were deposited by sinuous meandering streams that carried high loads of suspended rock particles. Both units include sandstones that originated as sand bars in stream channels, along with fine-grained flood-plain deposits of mudstones and siltstones laid down when streams overflowed their banks.

Bedding in the Dockum strata was little disturbed by vegetation, suggesting that the strata were either deposited too quickly for vegetation to develop, or that conditions were too arid. Vegetation may have grown only along streams where water was available year round.

Although petrified wood is found in places in some stream channel and lake deposits, few large tree trunks are found in the Dockum, in contrast to the Chinle Formation, its equivalent in Arizona. However, the latter contains abundant volcanic ash, which may have helped silicify the logs and contributed to their preservation while the Dockum Group has little volcanic ash.

Right *This reconstruction of the Middle Triassic, the time of the Tecovas Formation in Palo Duro Canyon, shows the Ouachita-Marathon Mountains in south Texas, continuing to the northeast where they become the Appalachians. The Gulf of Mexico was beginning to open south of Texas and the Atlantic Ocean was in the early stages of its formation along the Appalachians. Note that Texas was much closer to the west coast than it is today.*

Diagram by R.C. Blakey, website vishnu.glg.nau.edu/paleogeogwus.htm.

Right *Not much changed by the late Triassic, the time of the Trujillo Formation. The Gulf of Mexico and the Atlantic Ocean were a little more developed, and the Ouachita-Marathon and Appalachian Mountains were uplifted. A river, called the Chinle River by geologists, flowed from the Ouachita-Marathon Mountains northwest across Texas, New Mexico and Utah to the west coast.*

Diagram by R.C. Blakey, website vishnu.glg.nau.edu/paleogeogwus.htm.

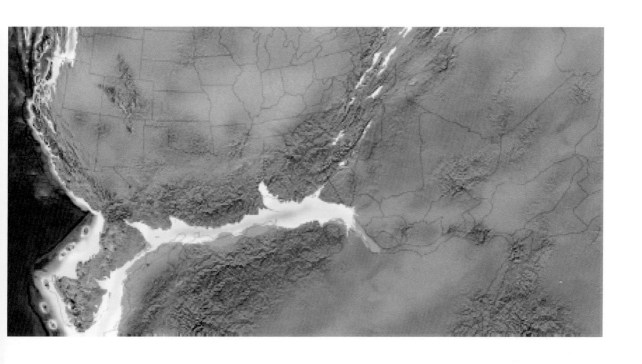

This photograph of the Trujillo sandstones was taken from the turn to the left on the road down into the canyon. On the far right of the photograph, a substantial thickness of Blackwater Draw strata overlies the cream colored Caprock. The Ogallala Formation below the Caprock is mostly obscured by rubble.

The Trujillo Formation

Descending into Palo Duro Canyon, the beautiful Triassic Trujillo sandstone cliffs come into view on the right just as the road makes a hard turn left (photograph on the previous two pages). The Triassic period was relatively short, lasting from 251 to 206 Ma (see chart on page 15) and the Trujillo Formation, and the one below it, the Tecovas Formation, belong to the later half of the period, from about 224-216 Ma. In this area of the Park, Trujillo sandstones lie directly below Ogallala sands and silts, at a slight angle to them. Geologists call the boundary between the two an *unconformity*, a break or gap in the geological record. The Trujillo unit is 210 million years older than the rocks above it. At one time there may well have been Jurassic, Cretaceous or early Tertiary rocks between the two but if so, they have long since been eroded away.

The Trujillo Formation consists of three sandstone beds up to 60 feet thick, separated by thin layers of red and gray shales. In some

Right *The full Trujillo section, 200 feet thick, is seen in this photograph, taken near the base of the road down into the canyon.*

The topmost red shale bed provided the iron oxide that stained the underlying sandstones.

Left *This beautifully sculpted Trujillo sandstone crops out as you turn right at the top of the long incline into the canyon.*

The sandstone, which is quite soft, has been eroded by wind.

places in the Park, the top bed in the formation is a red shale, as seen in the photograph below. The sandstones are composed of medium to coarse quartz and mica grains in a calcium carbonate matrix with conglomerate beds occurring in places. Weathered surfaces are often stained red or dark brown by iron oxides; fresh surfaces are gray or greenish gray.

The red or gray shales are flood-plain deposits left behind as fine-grained silt and clay that, in time, became shale. The maroon or red color comes from the mineral hematite (iron oxide, chemical composition FeO) in the rocks which is red if in fine particles.

The incline continues through the remainder of the Trujillo Formation as seen in the photograph below. The thickness of the Trujillo Formation in the Park when complete averages about 200 feet. However, the lowest of the three Trujillo sandstones is often the only part of the formation remaining on many of the buttes and mesas within the canyon, the others having disappeared through erosion.

The Upper Canyon

The pullout halfway down the incline into the canyon is a good place to view the upper canyon.

The line of light-green cottonwood trees marks the bed of Prairie Dog Town Creek.

The escarpment at right displays the reddish-brown Caprock near the top, the Trujillo sandstones, the Tecovas sandstone forming a ledge about halfway down, the yellow, gray and lavender Tecovas shales and the brick-red Quartermaster shales at base.

The Pioneer Amphitheater

This view of the Pioneer Amphitheater is from the Triassic Trail on top of Goodnight Peak. The theater is the site of an annual musical entertainment, "Texas", produced by the non-profit Texas Panhandle Heritage Foundation, founded in 1961. A cast of 65 actors, singers and dancers puts a show on five nights a week from early June to mid-August each year.

The show, written by Paul Green in 1966, dramatizes the determination and pioneering spirit of life in the Texas Panhandle in the 1880s. It includes galloping horses and special effects such as fireworks and simulated thunder and lightning. An attendance of 65,000 is expected for 2007. For more details visit www.texas-show.com.

As well as enjoying the pageant, you can have a steak dinner on the covered patio next to the amphitheater, take a backstage tour and shop in the gift shop.

Photo courtesy of www.texashiking.com.

Goodnight Peak

Goodnight Peak behind the Amphitheater shows off the full variety of the Tecovas Formation from its base yellow and lavender shales seen at mid-left, the light-gray sandstone bed above stained by iron oxide, and the maroon shales below the lower Trujillo sandstone capstone of the peak. Brick-red and white beds of the Quartermaster Formation crop out near the base of the escarpment.

The Tecovas Formation

The lower 50 feet of the Tecovas Formation consists of a series of mudstone beds, yellow, gray and lavender in color. These are overbank deposits, created when streams overflowed their banks and left clay and silt on the plain. The colors change from bed to bed as the level of oxygen at the time of deposition changed. Well-oxidized layers are colored red by the red iron oxide hematite. Yellow and gray beds contain poorly-oxidized iron oxide, magnetite or similar minerals, and probably represent material deposited in stagnant water. A few feet of these beds are exposed above Quartermaster strata, just before the first hairpin turn at the bottom of the incline, as shown in the photograph above.

Overlying the multicolored mudstones is a sandstone, 40 feet thick, white, crumbly, and known as "sugar stone" because of its even medium-grained sand particles. It can be distinguished from the overlying Trujillo sandstones by its lighter color, and absence of flakes of mica. Like the Trujillo sandstones, it is a weakly-cemented channel deposit. Channel deposits, because of their nature, vary in thickness over short distances, and the average thickness in the Park of Tecovas sandstone is said to be only 15 feet. Incidentally, no Triassic sandstones appear in the Canadian River breaks at Amarillo; the yellow mudstones are much more prominent and gave Amarillo its name.

The upper Tecovas consists of a maroon shale bed, 50 feet thick on Triassic Peak. In some parts of the Park, its color changes to red. In other places, it includes green beds. The full Tecovas section can be seen on Triassic Peak ahead (pages 62-63).

Left *The only exposure of the Tecovas Formation coming down the incline into the canyon is on the cliff above the first hairpin bend near the base of the cliff.*

Here, several of the base yellow and lavender mudstones can be seen above red and white Quartermaster shales.

Triassic Fossils

The Tecovas and Trujillo Formations were created in the latter half of the Triassic period, from about 224-216 Ma. This, of course, is a tremendous step back in time from the fossils last discussed, those of the Cita Canyon Lake Beds, which were 4 or 5 million years old at most. The Triassic period was one third of the way back to the beginning of life as we know it, to the world of reptiles.

I have already discussed the mass extinction that defines the Permian-Triassic boundary. The boundary itself is missing in Palo Duro Canyon although it may be present quite close by. For example, in Caprocks Canyons State Park thirty miles to the south, a volcanic ash bed in the Dewey Lake Formation at the base of the Caprock Escarpment is of an age that puts it in the very early Triassic period, just at the boundary.

In Palo Duro Canyon, the Tecovas Formation is about 35 million years younger than the Permian Quartermaster Formation below. Pangea had begun to break apart, the shallow salt seas that had overlain many continents had withdrawn, and non-marine Triassic red beds began to be deposited in many areas around the globe. Pangea lay across the equator and the climate was relatively uniform from the equator to high latitudes.

Life had begun to diversify again; it was an extraordinary time in the evolution of the predominant animals, tetrapods. Tetrapods (four-legged in Ancient Greek) are four-legged vertebrates, and include amphibians, lizards, crocodiles, dinosaurs, birds and mammals. Most of these groups first appeared during the Upper Triassic.

In the Triassic rocks of Palo Duro Canyon, vertebrate fossils generally occur in pockets in the mudstones of the Tecovas Formation. Many of the best vertebrate fossil sites are found in or near lake deposits, including one in Little Sunday Canyon, about 4 miles south of the Interpretive Center.

Triassic Peak

The best view of the Tecovas Formation in the canyon is from the rise beyond the Palo Duro Riding Stables. Tecovas strata crop out between the red and white Quartermaster rocks at its base and the gray Trujillo sandstone on its summit.

At the base of the Tecovas is a 15-foot light gray and lavender sandstone bed. Above it is a series of gray, yellow and lavender mudstone beds, 50 feet thick. These are overbank deposits, created when streams overflowed their banks and left clay and silt on the plain between streams.

Above the multicolored mudstones is a 40-foot sandstone, white and crumbly, known as "sugar sand" because of its even medium-grained sand grains. It can be distinguished from the overlying darker Trujillo sandstones by color, and lack of mica grains. This sandstone, like those in the Trujillo Formation, is a weakly-cemented channel deposit.

At the top of the Tecovas is a maroon shale bed, 50 feet thick here. In some parts of the Park, its color changes to red and in places, it includes some green beds.

Triassic Peak has a 15-foot Trujillo sandstone remnant at its peak. This bed caps a number of the lower mesas and buttes along Park Road 5, including Sorenson Point above the Hackberry Camp Area (page 66).

Desmatosuchus

An artist's impression of a *Desmatosuchus*, a reptile with an armored body and a pig-like head. The animal grew up to 17 feet long and weighed up to 2,000 lbs. It had a pair of huge, backwardly-turned shoulder horns, preceded by several pairs of smaller spikes on the side of the neck.

Desmatosuchus was a predator that fed on fish and small animals and lived mostly in water. The head was large and flat, with the eyes looking upwards. It probably spent a great deal of time submerged and motionless, waiting for prey. *Desmatosuchus* was a strong swimmer, but would have been very clumsy on land, and it is likely that it ventured from water rarely, if at all.

The most common fossils in the Tecovas are the teeth and bony plates of *Phytosaurs*, large semi-aquatic predatory reptiles, 6 to 40 feet long. They were very similar to modern crocodiles in size, appearance, and lifestyle, and in fact shared a common ancestor with crocodiles. However, crocodiles only appeared near the end of the Jurassic period, 50 million years after *Phytosaurs* became extinct.

Another common fossil found in the Tecovas is *Metoposaurus*, a large flat-headed amphibian up to 10 feet long. It has small, weak limbs that denote its mostly water-based life. Its main diet was fish which it captured with its wide jaws lined with needle teeth. *Metoposaurus* fossils found near Amarillo in 1940 are displayed in the Panhandle-Plains Historical Museum in Canyon.

Remains of *Desmatosuchus*, a crocodile-like amphibian with an armored body and a pig-like head, have been found in Big Sunday Canyon.

Fish remains are found from time to time in the Tecovas, mainly the teeth of sharks, ray-finned fish, lungfish, and coelacanths. The latter, closely related to the lungfish, were believed to have become extinct at the end of the Cretaceous period, about 65 million years ago, until a live specimen was found off the east coast of South Africa in 1938. Since then, live specimens have been found at several sites in the Indian Ocean. Lungfish can breathe air and are able to bury deeply into mud in dry periods and reduce their metabolism to a sixtieth of its normal rate. Their presence here suggests that streams in the late Triassic period dried up from time to time.

Although dinosaurs had developed by this time, no dinosaur fossils have been found in the canyon. Fossil remains of a dinosaur called *Coelophysis*, the earliest known dinosaur, have been found in the upper Trujillo Formation in nearby Garza County.

Coelophysis - The Oldest North American Dinosaur

Skeleton drawing of a *Coelophysis*, the oldest dinosaur found in North America. The adult was between 6 and 9 feet long and measured 6 feet high at the shoulders. The animal's teeth show that it was carnivorous. The name means "hollow form", the animal having hollow limb bones. The adult weighed only about 66 lbs and was fast moving.

Several hundred *Coelophysis* fossils, many of them complete skeletons, were found at Ghost Ranch in New Mexico in 1947. It is thought that the herd was buried by a flash flood so quickly that their bodies were protected from scavengers.

The Quartermaster Formation

The Hackberry Camp Area provides the first opportunity to see Permian rocks of the Quartermaster Formation up close. Two Permian formations crop out in the Park. The upper one is the Quartermaster Formation, about 60 feet thick on average, which forms the floor and lower walls of the canyon. Underneath it, the Cloud Chief Formation only appears in a few locations in the Park, but about a mile downstream from the turnaround, the creek cuts through the base of the Quartermaster Formation and from there downstream the Cloud Chief Formation is widely exposed (see regional map on page 18).

The Quartermaster Formation consists of brick-red or vermilion shale with lenses of gray shales, clays, mudstones and sandstones, and veins, lenses and beds of gypsum. The strata were laid down in coastal tidal flats during the final retreat of the Permian ocean; some of their features appear to have been created by fresh-water streams running across the flats as the ocean advanced and retreated. The formation belongs near the end of the Permian period, perhaps about 254 million years ago. It contains no fossils so its exact age is unknown.

The Cloud Chief Formation

The Cloud Chief Formation, as shown on the map on page 20, crops out in the Park where Sunday Creek runs into the Prairie Dog Town Fork, in a small area called by geologists the Velloso Dome, and at three other small exposures farther down the Prairie Dog Town Creek. The outcrops consist of about 15 feet of gypsum layers from 1-24 inches thick interspersed with thin lenses of mudstone up to one inch thick.

At the Velloso Dome, the Quartermaster shales appear to have been arched up from below, and the Cloud Chief Formation is nearly 200 feet above the position where it begins cropping out continuously about four miles downstream. How does one account for this abnormality? Two explanations have been put forward. One is that, while the area has continually subsided as Permian evaporites have been dissolved, perhaps the Velloso Dome is a place where this has not happened. That explanation seems unlikely.

The second explanation is rather more complicated. In the late Permian, gypsum precipitated from coastal waters as they dried up, and was later transformed into anhydrite beds by heat and pressure over time. (Anhydrite ($CaSO_4$) is a dehydrated form of gypsum ($Ca(H_2O)_2(SO_4)$)). In recent times, the explanation goes on, groundwater recombined with the anhydrite to reconstitute gypsum, and as gypsum takes up 30 to 50 per cent more volume than anhydrite, the rock expanded, creating doming.

Left *A good example of the lower succession in the canyon can be seen behind the Hackberry Camp Area.*

The lower part of the escarpment has red Quartermaster shales at base with multi-colored Tecovas mudstones above. The summit, Sorenson Point, is capped by the Tecovas sugar sandstone.

Right *Cloud Chief Formation shales and gypsum beds crop out in the Velloso Dome, near the confluence of Sunday Creek and Prairie Dog Town Creek. The creek can be seen in the bottom right corner of the photograph.*

Photograph courtesy of Chuck Hassell

The Lone Star Interpretive Theater

A recent addition to the Hackberry Camp Area is the Lone Star Interpretive Center, dedicated in 2000. The site is used by Park staff for presentations on topics such as wildlife and history of the canyon to school and other groups.

The setting is stunning in the evening sun. Tecovas sugar sandstone is the caprock on Sorenson Point behind the theater with lavender and yellow Tecovas mudstones below, followed by the Quartermaster shales in the lower half of the escarpment.

Caves in Palo Duro Canyon

Most caves in Texas are found in limestone, created by rock dissolving in percolating groundwater. In the Quartermaster Formation, however, although the gypsum is soluble, the rock as a whole is not, and caves develop in it by "piping". In this process, individual grains of rock are removed one by one by percolating water, beginning at root channels at the surface and emerging at the foot of the slope through openings that become enlarged as time goes on.

In this example, the foot of the slope was evidently higher when the cave developed than it is now, the canyon floor having been lowered by erosion over time.

Hoodoos

Irregular rock pillars, called hoodoos in the southwest, develop in regions of sporadic heavy rainfall from rocks that have different resistance to erosion. Sandstone, for example, is more resistant to erosion by rain or wind than mudstone or shale, so an outcrop will erode down to a sandstone bed. Softer beds under the sandstone are protected to some extent by the sandstone cap and form pillars.

The results can be seen in numerous places around the Park, some quite large, some small, as the three examples on these pages demonstrate. A hoodoo eventually disappear as the underlying pillar thins. The Devil's Tombstone, for example, which had been a well-known landmark in the Park since its opening, recently collapsed.

Above *Hoodoos can even develop on jumbled rock falls, as seen here, to the right of the Pioneer Amphitheater parking lot. A small sandstone slab has protected the underlying Tecovas mudstones and Quartermaster shales from erosion, creating a miniature hoodoo.*

Left *The Lighthouse, a photographer's favorite, is a good example of a hoodoo. Its cap is a slab of Trujillo sandstone, underlain by Tecovas mudstones and thin sandstones.*

Photograph © Texas Parks and Wildlife Department.

Right: *A hoodoo on the south end on Capitol Peak is seen here in close-up. It has a pillar of Tecovas mudstones with thin sandstone beds on a base of Quartermaster shales.*

The lower sandstone bed seems to have slumped down to the left. The head of the figure, another sandstone layer, is perched precariously on top. This hoodoo is quite likely to collapse in the near future.

Capitol Peak

Panoramic view of Capitol Peak, taken from the Lighthouse trail. The hoodoo on left was shown in close-up on the previous page.

The Tecovas Formation here has a sandstone bed at base, between the Quartermaster shales and the Tecovas yellow and gray mudstones.

Wildflowers in the Park

The canyon is carpeted by wildflowers in the spring, especially after a wet spring such as 2007, when these photographs were taken.

Above The pencil cholla or tasajillo, *Opuntia leptocaulis*, is particularly colorful when its red berries have developed.

Right above Bladderpod, *Lesquerella fendleri,* a member of the mustard family, carpets the area with yellow flowers in the spring.

Right below Feather dalea or limoncillo, *Dalea formoso,* is a West Texas native.

PALO DURO VISTAS

Trujillo Sandstone Cliffs

These cliffs, photographed from the Goodnight Dugout parking area, provide an attractive view of the Trujillo sandstone at top with glimpses of Tecovas mudstone half-way down and red and white Quartermaster shale just above the tree-line.

See pages 96-97 for a description and photograph of a replica of the Goodnight Dugout.

Slumping in the Quartermaster Shales

Slumping in the Quartermaster, seen here looking south across the canyon to the Mesquite Park escarpment, is the result of dissolution by groundwater of evaporites in the underlying Permian strata.

In other places, such as the Velloso Dome, expansion of the underlying strata causes doming of the Quartermaster shales.

The Rock Garden

In this photograph, taken 500 yards beyond the fifth water crossing, a large outcrop of landslide and slump debris comes down to the road. These Trujilllo sandstone blocks are known locally as the Rock Garden.

Cow Camp Cabins

The four Cow Camp Cabins were built by the Civilian Conservation Corps in 1933-37 and restored in 2001. The cabins have two single beds and a fireplace, but no plumbing. Here, Gary Bennett and Mark Hassell of the Texas Parks and Wildlife Department are touching up one of the cabins in time for the summer rush.

Looking up the canyon from the Cow Camp Cabins

The Cow Camp Cabins are built on a ledge of Quartermaster shale, 20 feet above the stream bed of Prairie Dog Town Creek. This photograph looks upstream from the ledge. The light green cottonwoods delineate the creek bed.

Left *This cottontail rabbit was photographed near the Cow Camp Cabins.*

Below *These wild turkeys at the Lone Star Interpretive Theater, are quite tame from being fed by campers.*

Maroon Trujillo Cliff

A thick bed of Caprock separating Blackwater Draw from Ogallala strata can be seen in this photograph taken from the turnaround. The upper maroon unit of the Tecovas Formation is well exposed above the lower slopes of Tecovas multicolored mudstone.

You may have noticed that the mudstones are almost always sheathed with a coating on which no plants will grow. The sheathing is composed mainly of calcite, the mineral form of calcium carbonate. The mudstones have a high calcite content which is carried to the surface by water seeping out from the interior.

These slopes are slippery, steep and very dangerous. It is impossible to get a hand or foothold once you start sliding. The rule of thumb is to avoid slopes on which no vegetation grows.

Mesquite Park Cliffs above the Turnaround

The even cliffs that bound Mesquite Park southeast of the turnaround are seen in this photograph taken just beyond the turnaround. The full succession is on view, from the Blackwater Draw sands and silts at top to the Quartermaster shales at base. The top of the cliff is 625 feet above the Prairie Dog Town Creek.

Fortress Cliff

Previous page You can see why it is called Fortress Cliff, seen here in the late afternoon from the south side of the canyon. The entire upper succession is on view, beginning with light-colored Blackwater Draw sediments above the even light brown Ogallala Formation.

Next, are gray Trujillo sandstones, stained red in places, the thick maroon upper Tecovas unit just above the vegetation and the gray Tecovas sandstone bed.

The remainder of the escarpment is covered by slump debris, including a large block of Quartermaster strata near the right of the photograph.

Tecovas yellow and lavender mudstones overlie Quartermaster shales in the right foreground. At bottom, light green cottonwoods indicate the stream channel of Prairie Dog Town Creek.

Quartermaster Formation Gypsum Veins

This close-up view of a Quartermaster bank on Alternate Park Road 5 near the Mesquite Camp Area shows thin gypsum beds and veins in the shale. The veins show that some of the gypsum flowed under pressure and recrystallized in fractures in the shale.

Three varieties of gypsum are present in the Quartermaster, satin spar, a white, translucent mineral with a silky luster, selenite, a colorless variety that often forms sheets, and alabaster, a hard, very fine-grained white massive material. Most of the veining is satin spar.

The Devil's Slide

This sharp spur of Tecovas and Quartermaster strata is distinct enough to be seen on satellite photographs of the area. The lower 50 feet of the Tecovas Formation consists of a series of shale or mudstone beds, gray, yellow and lavender in color. These are overbank deposits, created when streams overflowed their banks and left clay and silt on the ground between streams. The different colors result from different conditions at the time of deposition. Lavender material contains hematite and was most likely deposited on dry land. Yellow and gray beds were deposited under water, perhaps in lakes.

As I mentioned on page 82, these slopes are very dangerous. Is is impossible to gain hand or footholds once you start sliding. Several accidents here have led to the name, the Devil's Slide.

Quartermaster Shale

Another fine exposure of the lower strata from Alternate Park Road 5 on the way back to the park exit.

The caprock is once again the lower Trujillo sandstone with maroon upper Tecovas shale below and then Tecovas sugar sandstone capping ledges on the left and right. As on Capitol Peak, a sandstone bed separates the yellow and gray mudstones from the Quartermaster shales.

Exploration of the Llano Estacado

Palo Duro Canyon penetrates deeper into the Llano Estacado than any other canyon and was thus one of the most hospitable sites for early man, providing shelter, wood, water and plant and animal life. The first traces of human activity in the area date from near the end of the Ice Ages (1.8 million years ago to 10,000 years ago).

Early humans were hunters who used spears and lances to kill animals. Their activities can be dated by the type of spear points they used. Remains of broken points have been found in many places along the canyon rim. The hunters probably sat up on the rim to look out for game and replaced their broken points while they waited.

The northern hemisphere climate changed abruptly about 11,700 years ago and many North American animals became extinct (see page 38). Almost no human traces have been found for the period between the extinction and about 5,800 years ago, when hearth stones, flint flakes, bone fragments, mortar holes and graves became common. Hunters at that time pursued the modern bison and in the final prehistoric period, grew crops and produced pottery.

Europeans arrived on the Panhandle in 1541 with the expedition of the Spaniard Francisco Vásquez de Coronado. The Spaniards, who had found fabulous wealth of gold and silver in Mexico and Peru, had very little knowledge of the vast continent to their north but had no reason to suppose that it was any less endowed. A persistent myth in Spanish culture, indeed in all European culture, was that of the seven cities of gold, dating back to the conquest of Spain and Portugal by the Moors in the twelfth century. The legend had seven bishops of the Catholic Church and their followers escaping across the Atlantic to a land called Antila. There they founded the cities of Cibola and Quivira which grew enormously wealthy from gold and gemstones. Eventually the legend expanded to include seven cities, one for each bishop.

In 1536, Alvar Nunez Cabeza de Vaca arrived in Mexico City from the north. Sailing from Spain as treasurer of a 300-man expedition to Florida, Cabeza de Vaca, after many misadventures, was one of four survivors who came ashore near Galveston Island. Enslaved by natives for about six years, de Vaca and his companions eventually escaped, traveling across Texas and northern Mexico. Finally arriving in Mexico City, they brought with them rumors of wealthy cities to the north where people lived in large houses. Rumors immediately connected the description of wealthy cities to the north to the seven cities of gold.

To verify Cabeza de Vaca's statements, Viceroy Antonio de Mendoza sent Marcos de Niza to the north in the spring of 1539 with one of the Cabeza de Vaca party, a black slave called Esteban or

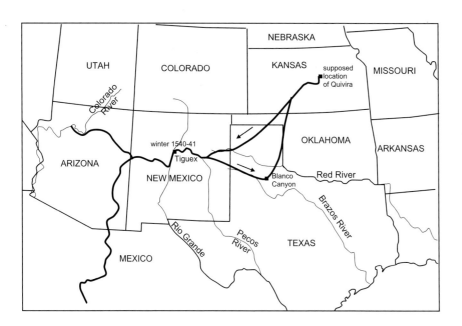

Left *The Coronado Expedition crossed the American West in 1540-41, and included a detour to the Grand Canyon. The route from Tiguex east is uncertain. Modern archeology points to Blanco Canyon near Plainview as a likely stopping point for the expedition. Coronado continued on to Kansas with 30 men before returning to join the main army at Tiguex. The return route from Quivira to Tiguex shown on the map is speculative.*

Estevanico, to guide him. On his return, Niza confirmed that he had seen a place called Cíbola but had not entered it for fear of his life, as Esteban had been killed there by the natives. He had seen Cibola from a distance and said that it was bigger than Mexico City.

Mendoza appointed Francisco Vázquez de Coronado, governor of the New Spain province of Nueva Galicia, to lead an expedition to conquer the area, funded by the Viceroy and Coronado's wife. It was a substantial operation; 335 soldiers, 1000 friendly Indians, 1000 horses, 500 cows, and over 5000 rams and ewes. The expedition first arrived in Arizona, visited the Grand Canyon, and then moved on to New Mexico. There, Coronado was greatly disappointed to find that there was no gold in the towns of Cibola, the present-day Zuni reservation of west central New Mexico. After spending the winter in Tiguex, on the Rio Grande near today's Albuquerque, they traveled east, following the suggestions of an Indian called El Turco who claimed he could lead them to Quivira.

Pedro de Castaneda of Najera, a soldier on the expedition, vividly described the expedition's adventures in his memoirs, written around 1560. As the army crossed the Llano Estacado, Coronado sent scouts ahead who reported finding a great *barranca*, a canyon like the *barrancas* of Colima, a mountainous area in southern Mexico, most likely Tule Canyon, about 20 miles south of Palo Duro Canyon.

The army camped in another large *barranca*, the site unknown until a rancher near the town of Floydada, east of Plainview, turned up a mailed gauntlet on his land in the 1960s. This led to a flurry of interest in the area. An amateur archeologist, Jimmy Owens, an expert in using metal detectors, found a variety of metal objects, including nails probably from horses' hooves, and most importantly, bolt heads from crossbows. The latter was key to dating these artifacts: only early Spanish soldiers had crossbows; later expeditions were armed with longbows.

The Floydada site, in Blanco Canyon, is now considered to be the site where Coronado's army camped for some time. In Castaneda's words:

> "Thus the army reached the last barranca, which extended a league from bank to bank *(a league is about 2.6 miles)*. A small river flowed at the bottom, and there was a small valley covered with trees, and with plenty of grapes, mulberries, and rose bushes. This is a fruit found in France and which is used to make verjuice. In this barranca we found it ripe. There were nuts, and also chickens of the variety found in New Spain, and quantities of plums like those of Castile."

Coronado picked a small, light contingent of 30 men to travel north to Quivira, sending the main army back to their base at Tiquex, where they arrived in June, 1541. In July, Coronado reached Quivira, in present-day Kansas. Nothing of any value could be seen. On being interrogated, El Turco admitted that he had been brought up in this area and had wished to return home. He was executed for treason. The expedition returned to New Spain in disgrace where Coronado died a broken man in 1554.

Incidentally, the term Llano Estacado is often attributed to the Coronado expedition although it is not found in the written records of the expedition. The geologist W.F. Cummins in his 1891 report on the geology of the area said that:

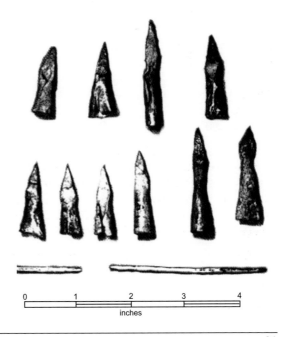

Right *Copper crossbow arrow points found at the campsite near Floydada, Texas. These and several dozen more were scattered over a football-field-sized area in a canyon floor, along with other Coronado-era material.*

> "It is suggested that the word from which our Staked Plains is derived is not the one that was originally used. That instead of Llano Estacado it ought to be Llano Estacada. Estacado is the perfect participle of estacar, which means staked plains. Estacada in the Spanish language means a palisade, and it is supposed that the term was used in reference to the Staked Plains in the accommodated sense in which we use the term palisade in the English language.
>
> It is supposed that the two words became confounded and changed at some later period, and that some one in attempting to explain the origin of the then used term estacado invented the theory of putting stakes across the Plains as guides."

This seems more probable i.e. the name should be translated as the "palisaded plains". The word palisade means a defensive stockade or a line of steep cliffs, usually along a river. Here, we have steep cliffs along the Canadian River and along the Caprock Escarpment. The title probably came into use by *comancheros* who came east from New Mexico to trade with the Comanches and *ciboleros* who came to hunt buffalo.

Early U.S. Explorers

It is hard to believe, as visitors to the Panhandle barrel along at 70 miles an hour towards Canyon, how fearsome an obstacle the High Plains were to the nineteenth century U.S. traveler. Zebulon Pike was the first of the Anglo-American explorers on the southern plains. In 1808 he wrote that the region would be as famous as the Sahara Desert due to its emptiness:

> "Our citizens will through necessity . . . leave the prairies [that are] incapable of cultivation to the wandering and uncivilized aborigines."

Twelve years later, U.S. Army explorer Lieutenant Stephen Long, in exploring the headwaters of the Red River, declared that the southern plains were

> "almost wholly unfit for cultivation and of course uninhabitable for people depending upon agriculture for their subsistence."

Long called it "The Great American Desert". Gradually, however, a succession of Army surveyors and intrepid frontiersmen found ways to cross it. One of the first Americans to do so was the trader Josiah Gregg. In 1840, he found a more direct route from Santa Fe to the Mississippi valley than the existing one though Saint Louis by following the Canadian River, fighting off Pawnees and blue northers as he went. The army topographical engineer James Abert followed Gregg's route in 1845 in a survey of the Canadian River.

However, nearly five years later, the regular infantry officer Captain Randolph Marcy gave his jaundiced view of the Llano in his widely-read "Route from Fort Smith to Santa Fe, Letter from the Secretary of War . . . February 21, 1850":

"When we were upon the high table land, a view presented itself as boundless as the ocean. Not a tree, shrub, or any other object either animate or inanimate, relieved the dreary monotony of the prospect. It was . . . the dreaded Llano Estacado . . . a land where no man, either savage or civilized, permanently abides . . . a treeless, desolate waste of uninhabited solitude, which always has been and must continue uninhabited forever."

By 1852, however, Marcy was back on the plains in command of a seventy-man exploring expedition across the Great Plains in search of the source of the Red River and directed to "collect and report everything that may be useful or interesting". He is said to have discovered the sources of both forks of the Red River, and became the first white man to explore the Palo Duro and Tule canyons.

Nevertheless, the High Plains deterred settlement by its inhospitable lack of water and trees, dramatic temperature changes, endless wind, harsh climate, and most especially by its Comanche inhabitants. It was not considered a place to settle; most people just passed through on their way to the richer territories of the Rocky Mountains and Pacific Coast.

The Comanches had arrived from the north in the early 1700s, and quickly drove out all other tribes. Fearsome warriors, expert horsemen, they had all the attributes of the Mongols, who had conquered Asia ands most of eastern Europe in the twelfth and thirteenth centuries, except that they had no Genghis Khan to forge them into a united people. Any young warrior could attract followers for hunting and raiding parties and create mayhem in Texas and northern Mexico.

In 1867, the Federal government created a reservation for the Comanches in Oklahoma just northeast of the Panhandle, but young braves found reservation life tame and periodically went hunting buffalo, their predominant food. However, after the Civil War, a fashion for buffalo coats in the eastern cities and the use of buffalo leather for industrial machinery belts and army boots in Europe made buffalo skins immensely valuable and led to massive killing of the buffalo herds on which the Comanche depended. Things came to a head in 1874 when a Comanche war party attacked a buffalo hunter camp at a place called Adobe Walls. The Federal Government organized an Army expedition to corral them.

In as many as 20 engagements, the Army drove the Comanche bands to the west throughout 1874. One of the final episodes took place in Palo Duro Canyon, when an army platoon under command of Captain Ranald McKenzie surprised a large Comanche encampment in the canyon, and captured and destroyed their horse herd. The campaign finally ended in June 1875 when the last Comanche band, Quanah Parker and his Quahadis, entered the reservation at Fort Sill.

The Goodnight-Loving Trail

This shadow relief map of the Goodnight-Loving trail shows the High Plains in their entirety, from southern South Dakota to Texas, where they blend into the Edwards Plateau. Palo Duro Canyon shows up in sharp relief east of Fort Sumner.

The Goodnight-Loving trail avoided the High Plains, and the Comanches and Kiowas who hunted buffalo there, by following the Middle Concho River to Horsehead Crossing (near today's Crane), then up the Pecos River valley to Fort Sumner and the Canadian River valley through Raton Pass into Colorado.

In later trails, Goodnight used Trinchera Pass, 20 miles west of Raton, to avoid paying a toll of 10c per head at Raton Pass.

In all, Goodnight trailed cattle for nine years, usually nine or ten thousand head per year, and established a ranch or "swing station" for grazing in Apishapa Canyon, 45 miles southeast of Pueblo.

The Coming of the Cattlemen

News of the Comanches' surrender spread fast, even to Colorado where the Texas cattleman Charles Goodnight had a ranch. Goodnight was one of those larger than life figures that abounded in nineteenth-century Texas. Born in Illinois in 1836, at age nine he rode bareback 800 miles with his family to settle in Milam County. At age eleven, his mother was widowed and Charles began working at neighboring farms and plantations, eventually graduating to freighting with ox teams. By the time he was fourteen, Charles was an accomplished Indian tracker, having learned the craft of tracking and hunting from an old Indian called Caddo Jake. When he was seventeen, his mother married again, and he and his stepbrother went into the cattle business, watching his stepbrother's brother-in-law's livestock, and receiving every fourth calf as payment. They gradually moved west, settling in the Keechie valley in Palo Pinto County, near today's Mineral Wells, and bringing their parents there in 1858.

Goodnight joined the Texas Rangers under another famed westerner, Jack Cureton, as a scout and guide. Among his many exploits, he led the rangers to a Comanche camp on the Pease River where they found an Anglo woman, Cynthia Ann Parker, who had been abducted from her family as a 10-year old in 1836. By a quirk of fate, her son Quanah Parker became a Comanche chief and later, Goodnight's close friend. On the outbreak of the Civil War, Cureton's rangers became part of the Frontier Regiment, guarding the western frontier, and Goodnight spent the Civil War years chasing outlaws and Indians along the frontier from the Canadian to the Colorado Rivers.

Post-Civil War Texas found the cattle business in ruins. The neglected herds had expanded enormously leading to ruinous beef prices, so Goodnight, who had managed to collect 180 head by onset of the Civil War, decided the only way to survive as a cattleman was to take livestock north to better markets, the Indian reservations, Colorado mining camps, and the Union Pacific railroad crews now working their way across Wyoming.

Goodnight partnered with an older cattleman, Oliver Loving, who had prospered during the war supplying beef to Confederate forces. Loving had earlier in his career trailed cattle to Colorado, so he knew the ropes. The first trail, in 1866, was with 2,000 cattle and 18 cowboys from Fort Belknap to Fort Sumner in New Mexico where a reservation had been established for Navajos and Mescalero Apaches. The subcontractors for the reservation bought their steers at the very high price of 8 cents a pound, leaving the partners a profit of $12,000, more than they ever dreamed of getting. Loving continued north to Colorado where he sold their cows and calves to the pioneer cattleman John Iliffe while Goodnight returned to Texas for more cattle.

Unfortunately poor Loving did not live long to enjoy his prosperity. On their third drive, he was shot while riding ahead of the herd in the Pecos River valley west of the Guadalupe Mountains and died of his wounds, after getting Goodnight to promise that he would bring his body back to Texas. Goodnight carried out his pledge the following summer, transporting the body in a metal casket made of oil cans that the cowboys had hammered flat, a famous episode borrowed by Larry McMurtry in his novel "Lonesome Dove".

Charles Goodnight continued for nine years in the business of exporting Texas cattle to feed troops and Indians in the west and to stock the northern territories of Wyoming and Montana. He married a girl from Weatherford, and established a ranch on the Apishapa River just east of Pueblo, Colorado. However, badly affected by the financial panic of 1873 (financial panics were a feature of late nineteenth-century America) which, along with overstocking of the range, soured him on Colorado, Goodnight decided to move back to Texas, with a view to settling below the Caprock.

Goodnight sent his wife to California until he was established and then drove his herd of 1,600 cattle into New Mexico where they wintered on the Canadian River in 1875. The next year, after returning to Pueblo to borrow money, he was led to Palo Duro Canyon by a Mexican mustanger (someone who captured wild horses), who told him of a wild gorge that cut the Llano in two and was the most wonderfully sheltered place for a ranch. After summering on the Canadian, the herd arrived at the edge of the canyon in November, 1876. Goodnight's own account of their entry into the canyon makes gripping reading:

Right *This replica of the dugout that Charles Goodnight used in his first winter in Palo Duro Canyon is on the right just beyond the second water crossing. Dugouts were widely used in the pioneer days of West Texas. Normally, they were partly dug into a bank, hence the name, and roofed with sods. They were easier to construct than normal housing on the treeless Plains and had good insulation from the heat and cold. However, they also had disadvantages: things dropped from the roof and the mud floor did not take kindly to rainy weather.*

> "Our first entrance into the Palo Duro Canyon was in November, 1876. We then made our entrance by way of the old Comanche trail between the junction of the Cañon Cito Blanco and the main Palo Duro Canyon. It took us about a half a day to work the cattle down this narrow and ragged trail. We then took the wagon to pieces and carried it down piece by piece on the mules. We had about six months' rations and much corn. This was also carried down on the mules. The canyon being narrow at this place prevented any buffaloes from being in there. Hence, grass and water were found there in abundance and the cattle ran at their own sweet will. The portage [getting supplies and cattle down in the canyon] was over after two days.
>
> We then started the herd down the canyon. As the canyon widened, the buffalo increased 'till finally by the time we arrived at the upper end of the wider valley (which has since been known as the "Old Home Ranch") we supposed we had ahead of us ten thousand buffaloes. Myself, Leigh R. Dyer and an Englishman named Hughes (son of the great Hughes of England) were the buffalo drivers.
>
> Such a sight was probably never seen before and certainly will never be seen again. The red dust arising in clouds, while the tramp of the buffalo made a great noise. The tremendous echo of the canyon, the uprooting and crashing of the scrub cedars made one of the grandest and most interesting sights that I have ever seen. If the buffaloes did

Left *Charles Goodnight in middle age.*

Photograph courtesy of the Panhandle Plains Historical Museum.

not come down off the mountain sides that were near us, we simply sent a sharpshooter ball among them. A nearby shot would cause an instant stampede, making kindling wood of the small cedars as they came.

These buffaloes were moved down the canyon some fifteen miles, giving ourselves room and grass for our sixteen hundred cattle. There we put on a herd line from the mouth of Turkey Creek and held them back, turning every day from eight hundred to fifteen hundred buffaloes."

Goodnight completed a dugout similar to the replica in the canyon and returned to Colorado to raise more funds and to meet his wife who had returned from California. After Goodnight was introduced to John Adair, an Irish financier who was interested in going into cattle ranching after seeing how profitable the business had become, he and Goodnight signed an agreement to set up a ranching operation financed by Adair and managed by Goodnight. This became the JA Ranch, still in existence and owned by descendants of Mrs. Adair, plus various other properties in the vicinity that eventually controlled over a million acres and more than 100,000 cattle.

Left *Charles Goodnight with Mrs. Adair. Although he often complained about John Adair's behavior, he was on good terms with Mrs. Adair.*

Photograph courtesy of the Panhandle-Plains Historical Museum.

One of the more interesting experiences of Goodnight's early life in the canyon took place in 1878 when one morning a runner came in to say that Comanches and Kiowas had come on to the ranch and were killing cattle in considerable numbers. They had tired of reservation life and had come looking for buffalo. Finding none, they began killing JA cattle.

Goodnight rode out to meet them. He found the Kiowas in an ugly mood, and avoided them, but followed the Comanches up the canyon proposing to their chief, Quanah Parker, that they meet to make a treaty. The next morning a dozen braves showed up and formed a circle with Goodnight and their interpreter in the middle.

After a good deal of hostile questioning, they came to an agreement. In Goodnight's words:

Statue of Charles Goodnight

The Charles Goodnight statue at the Panhandle-Plains Museum in Canyon, in front of a log cabin typical of the time, was donated by Mesa Petroleum, an oil and gas company with large holdings in the northern Panhandle.

The company was founded by Boone Pickens, who attended high school in Amarillo, and was headquartered there until 1981. It merged with Parker & Parsley Petroleum Company of Midland in 1997 to form Pioneer Natural Resources, headquartered in Irving, near Dallas. Boone Pickens retired as chairman of Mesa Petroleum in 1996.

'"What have you got to offer" they said.

'I've got plenty of guns and plenty of bullets, good men and good shots, but I don't want to fight unless you force me to". Then, pointing to Quanah, I said "You keep order and behave yourselves, protect my property and leave it alone, and I'll give two beeves every other day until you find out where the buffaloes are."'

The treaty held. In fact, Goodnight said he had never met an Indian who did not keep his word. To deal with the Kiowas, however, Goodnight had to ask the cavalry from Fort Elliot to come and intercede. They did and were able to negotiate a return by all bands to the reservation in the spring of 1879, their last large migration to the Panhandle.

Quanah Parker prospered in the reservation, becoming the leading Comanche spokesman and negotiator until his death in 1910. Much of the reservation's income came from leasing land to Texas cattlemen, although not to Goodnight.

When the original agreement with Adair expired in 1882, the partnership showed a profit of $312,000, and was extended for a further five years. John Adair died in 1885 but the partnership continued until 1887, Goodnight taking as his payment the part of the property known as the Quitaque Ranch, although he continued to manage the JA Ranch for Mrs. Adair until 1888. He sold the Quitaque Ranch the next year, bought 160 sections (102,400 acres) in Armstrong County and built a ranch house near the town that bears his name, moving there in December 1887 and severing his connection with Palo Duro Canyon.

Goodnight continued in the cattle business and became a renowned authority on cattle breeding. He also created a buffalo herd and experimented in cross-breeding buffalo with cattle, creating the *cattalo*. The descendants of his buffalo herd now form the herd maintained at Caprock Canyons State Park.

In later years Goodnight speculated in Mexican mining ventures and Colorado gold prospects. In 1929, at the age of 93, he died at his winter home in Phoenix and is buried in the cemetery at Goodnight, Texas. Looking back near the end of his long life, he said:

> "All in all, my years on the trail were the happiest I ever lived. There were many hardships and dangers, of course, that called on all a man had of endurance and bravery. But when all went well, there was no other life so pleasant. Most of the time we were solitary adventurers in a great land as fresh and new as a spring morning. And we were free, and full of the zest of darers."

Acknowledgments

With thanks to

Pat Dasch and Martha MacLeod for proof-reading the manuscript

Mark Hassell, Julius Dasch, Bernice Blasingame, and Robyn Prather for enthusiastically reading all or parts of the manuscript

Betsy Bustos, Panhandle-Plains Historical Museum, for retrieving archival photographs

Tom Lehman, Texas Tech University, for setting me straight on Permian and Triassic sedimentology

Don Dowdey, Sul Ross Library, for coding the Publishers Cataloging in Publication Data

Ron Blakey for the images of Permian and Triassic paleogeography

Chuck Hassell for the photograph of the Cloud Chief Formation

Ron Barron for the photograph of the Amphitheater

Notes

Page

The chief sources of information on the geology of Palo Duro Canyon were: Mathews, W.A. III, 1969, The geologic story of Palo Duro Canyon, Guidebook 8, Bureau of Economic Geology, Austin, 51 p.; Hood, H.C. and J.R. Underwood, 1979, Geology of Palo Duro Canyon in Guy, D. ed., *The story of Palo Duro Canyon*, Texas Tech Press, Lubbock, p. 3-35; and Guidebook of Palo Duro Canyon, 2001, West Texas A&M University Geological Society, 20 p.

The chief source of information on the paleontology of Palo Duro Canyon was: Wright, R.A., 1979, The paleontology of Palo Duro Canyon, *in* Guy, D. ed., *The story of Palo Duro Canyon*, Texas Tech Press, Lubbock, p. 59-117 and for Triassic paleontology: Lehman, T. and S. Chatterjee, 2005, Depositional setting and vertebrate biostratigraphy of the Triassic Dockum Group of Texas, J. Earth Syst. Sci. 114, No. 3, June 2005, pp. 325–351

10 The physiographic map of the Great Plains is based on: one by the U.S. Geological Survey available on the internet at http://tapestry.usgs.gov/physiogr/physio.html

12 The base shaded relief map is from: http://seamless.usgs.gov/website/seamless

13 The base shaded relief map is from: http://seamless.usgs.gov/website/seamless

13 The Socorro Fracture Zone is discussed in: Sanford, A.R. And Kuo-Wan Lin, 1998, Evidence for a 1400 km long Socorro Fracture Zone, Geophysics Open-File Report 89, Earth and Environmental Science Department, New Mexico Institute of Mining and Technology, Socorro, New Mexico 87801 (http://www.ees.nmt.edu/Geop/nmquakes/R89/R89.HTM).

14 The contour map is from: Gustavson, T.C. and R.J. Finley, 1985, Late Cenozoic geomorphic evolution of the Texas Panhandle and northeastern New Mexico - Case studies of structural controls on regional drainage development, Report of Investigations No. 148, Bureau of Economic Geology, Austin.

17 The nearest Permian salt layer is 790 feet below the surface at Canyon: see Gustavson T.C. and W.W.Simpkins, 1989, Geomorphic processes and rates of retreat affecting the Caprock Escarpment, Texas Panhandle, Report of Investigations No. 180, Bureau of Economic Geology, Austin, fig. 12 Structure-contour map of the top of the Upper Permian Alibates Formation.

18 The regional geological map is a composite from: Geologic atlas of Texas, Amarillo Sheet, scale 1:250,000, 1969 reprinted 1981 with limited revisions, and Geological atlas of Texas, Plainview Sheet, scale 1:250,000, 1968, revised 1992, Bureau of Economic Geology, Austin.

20 The State Park geologic map is based on: Geologic map of Fortress Cliff Quadrangle, Texas, Bureau of Economic Geology open-file map, 1998, Bureau of Economic Geology, Austin.

22 Playas are discussed in Gustavson, T.C., V.T. Holliday and S.D. Hovorka, 1995, Origin and development of playa basins, sources of recharge to the Ogallala aquifer, southern High Plains, Texas and New Mexico, Report of Investigations No. 229, Bureau of Economic Geology, Austin.

23 Data on processes and speed of canyon expansion comes from Gustavson & Simpkins, *op. cit.*

26 The early history of the State Park is based on: Petersen, P.L., 1979, A park for the Panhandle: The acquisition and development of Palo Duro Canyon State Park *in* Guy, D. ed., *The story of Palo Duro Canyon*, Texas Tech Press, Lubbock, p. 118-145.

33 The main source for general information on the Blackwater Draw Formation is: Gustavson, T.C., 1996, Fluvial and eolian depositional systems, paleosols, and paleoclimate of the upper Cenozoic Ogallala and Blackwater Draw Formations, Southern High Plains, Texas and New Mexico, Report of Investigations No. 239, Bureau of Economic Geology, Austin.

33 Map from Gustavson (1996) p. 6.

39 The main source for general information on the Ogallala Formation is Gustavson (1996), *op. cit.*

40 Original distribution of the Ogallala Formation from: Heller, P.L., K. Dueker and M. E. McMillan, 2003, Post-Paleozoic alluvial gravel transport as evidence of continental tilting in the U.S. Cordillera, Geological Society of America Bulletin, **115**, p. 1122–1132.

47 Permian and Triassic fossils separated by 23 to 33 feet of rock: Collinson, J.W., W.R. Hammer, R.A. Askin and D.H. Elliot, 2006, Permian-Triassic boundary in the central Transantarctic Mountains, Antarctica: Geological Society of America Bulletin, **118**, p. 747-763.

48 Permian Basin map is by the Bureau of Economic Geology, Austin.

50 The main source for Triassic sedimentary data is Lehman & Chatterjee, *op.cit.*

61 Dating of the Dewey Lake Formation in Caprock Canyons State Park: Steiner, M., 2001, Magnetostratigraphic correlation and dating of West Texas and New Mexico Late Permian strata, p. 59-68, *in* Lucas, S., and Ulmer-Scholle, D.S., ed., Geology of the Llano Estacada, New Mexico Geological Society Guidebook, v. 52.

89 Exploration of the Llano Estacado is recounted in great detail in: Morris, J.M., 1997, *El Llano Estacado,* Texas State Historical Association, Austin, pp. 414.

90 The full text of Pedro de Castanado's report is available on the internet site for the PBS Series "The West" at http://www.pbs.org/weta/thewest/resources/archives/one/. Several other documents on the Coronado expedition are also provided there.

92 The quote on the Llano Estacado name comes from: Dumble, E.T., 1892, Third annual report of the Geological Survey of Texas, Austin, page b129a, (http://www.lib.utexas.edu/books/landscapes/publications/txu-oclc-5235917-3/txu-oclc-5235917-3-b129a.html)

93 The theory on the extinction of the buffalo: Taylor, M,.S., 2007, Buffalo Hunt: International Trade and the Virtual Extinction of the North American Bison, NBER Working Papers 12969, National Bureau of Economic Research, Inc., available on the internet at http://www.econ.ucalgary.ca/fac-files/st/w12969.pdf

94 The shaded relief base map is from: http://www.shadedrelief.com/physical/usa_zoom.htm

95 The two main sources used on Charles Goodnight's life are: Haley, J. Evetts, 1936, *Charles Goodnight Cowman and Plainsman;* University of Oklahoma Press, Norman, Oklahoma, 485 p.; and Burton, Harley True, 1927, "History of the J A Ranch", Volume 32, Number 1, Southwestern Historical Quarterly Online, http://www.tsha.utexas.edu/publications/journals/shq/online/v032/n1/contrib_DIVL717.html and succeeding volumes.

100 Goodnight quote on the trail: Haley p. 259.